BASIC ASPECTS OF MEDICAL AND DENTAL LASERS

JEFFREY G. MANNI

Copyright © 2004, 2007, 2013 JGM Associates, Inc. (Jeffrey G. Manni).

All rights reserved. No part of this book may be reproduced, stored, or transmitted by any means—whether auditory, graphic, mechanical, or electronic—without written permission of both publisher and author, except in the case of brief excerpts used in critical articles and reviews. Unauthorized reproduction of any part of this work is illegal and is punishable by law.

ISBN: 978-1-4834-0266-6 (sc)
ISBN: 978-1-4834-0265-9 (e)

Because of the dynamic nature of the Internet, any web addresses or links contained in this book may have changed since publication and may no longer be valid. The views expressed in this work are solely those of the author and do not necessarily reflect the views of the publisher, and the publisher hereby disclaims any responsibility for them.

Any people depicted in stock imagery provided by Thinkstock are models, and such images are being used for illustrative purposes only. Certain stock imagery © Thinkstock.

Lulu Publishing Services rev. date: 07/24/2013

Please read this disclaimer . . .

This book is intended for informational and educational purposes only. JGM Associates, Inc. does not endorse or recommend specific companies, products, or uses of products. In all cases, practitioners who use lasers should complete an accredited training course, approved by the manufacturer, before attempting to use any laser product or related accessory. The reader should note that not all products have FDA clearance or approval for the same clinical uses.

Information has been obtained from sources believed to be reliable. While JGM Associates believes the data provided herein to be accurate, no expressed or implied guarantees are made regarding accuracy or adequacy. Product related data can change very quickly and without notice. Interested readers are encouraged to contact manufacturers directly for their latest product information.

This report is not intended as a substitute for proper due diligence. Although an attempt has been made to provide thorough coverage of the subject matter of this report, no guarantees regarding completeness are made. JGM Associates is not responsible for any results obtained from use of information provided herein.

JGM Associates, Inc. reserves all rights to this copyrighted document. This report may not be reproduced in part, or in whole, without written consent of an authorized officer of JGM Associates, Inc.

Table of Contents

Section 1: Medical Lasers

Laser energy and biomedical work 1
Lasers as energy and power conversion devices 2
Basic components of a laser 4
Solid-state, liquid dye, and gas lasers 6
Optical pumping 7
Solid-state laser media 8
Laser wavelengths and wavelength versatility 12
Nonlinear wavelength conversion 14
Laser beams 15
Temporal emission modes: pulsed vs. continuous-wave (CW) lasers 18
Pulse duration, pulse rate, peak power, average power 19
Quasi-CW operation: single pulse, chopped, super-pulsed 22
Pulsed modes: free-running, Q-switched, mode-locked 27
Semiconductor diode lasers 30
Diode-pumped solid-state lasers (DPSSL's) 32
Fiber lasers 34
Basic aspects of medical laser products 35

*Section 2: Laser-Tissue Interactions, Effects,
and Therapeutic Applications*

Laser therapy and procedures 1
Basic types of laser-tissue interactions 4
Laser irradiation of tissue 6
Absorption and absorption depth 7
Scattering 8

Penetration depth 9
Irradiated volume 9
Fluorescence 11
Photothermal interactions, effects, and therapy 11
    Laser hyperthermia 12
    Laser thermal keratoplasty 12
    Tissue welding 13
    Coagulation, necrosis, and hemostasis 13
    Vaporization and cutting 14
    Vaporization with pulsed vs. continuous-wave lasers 16
    Surgical precision and control 17
    Lasers as bloodless scalpels 20
    Selective photothermolysis 21
Photochemical interactions and ablation 22
Photochemical ablation 22
Photodynamic effect and photodynamic therapy (PDT) 23
Photoacoustic / photodisruption interactions, effects, and therapy 24
    Membranectomy 26
    Lithotripsy 26
    Ablation of calcified tissue 26
    Tumor disruption 27
Biostimulation and pain relief 27
Biomolecules, Tissues, and Materials 28
    Water 28
    Hemoglobin and blood 30
    Melanin 31
    Xanthophyll 31
    Carbon 31
    Collagen 33
    Soft tissue 33
    Blood vessels 33
    Cartilage 34
    Bone 34
    Dental tissue 35
    Atherosclerotic plaque 35
    Calculus 36

Section 3: Medical Laser Delivery Systems

Beam transfer hardware 2
    Articulated arms 2
    Optical fibers 5
    Hollow waveguides 7
Beam application devices 8
    Fiber handpieces 8
    Hollow waveguides 9
    Micromanipulators 10
    Hollow probes 11
    Arms vs. waveguides vs. fibers 11
    Scanning delivery systems 15
Visualization devices or modalities 16
    Endoscopes 16
    Surgical microscopes 18
    Other visualization / guidance methods 18

Section 4: Optical Fiber Delivery Systems

Bare fiber basics 1
Contact vs. non-contact use of bare fibers 5
Hot-tip fibers 6
Cool-tip fibers 7
Hot-tip vs. cool-tip fibers 7
Laser beam divergence, focusing, and fiber-coupling 9
Fiber numerical aperture (NA) 12
Infrared (IR) fibers 15
Photonic bandgap hollow waveguides 18

Section 5: Argon Lasers

Technical background—Argon lasers 1
Technical background—tissue interactions and effects (Argon lasers) 2
Technical background—considerations for intraoral use (Argon lasers) 3
Historical background (argon lasers) 4
Regulatory status and current clinical uses (Argon lasers) 4
Recent advances / developments (Argon lasers) 4

Section 6: Carbon Dioxide Lasers

Technical background—Carbon dioxide lasers 1
Technical background—tissue interactions and effects (Carbon dioxide lasers) 4
Technical background—considerations for intraoral use (Carbon dioxide lasers) 5
Historical background (Carbon dioxide lasers) 5
Regulatory status and current clinical uses (Carbon dioxide lasers) 6
Recent advances (Carbon dioxide lasers) 6
Product trends and possible future developments (Carbon dioxide lasers) 6

Section 7: Semiconductor Diode Lasers

Technical background—Diode lasers 1
Technical background—tissue interactions and effects (Diode lasers) 2
Technical background—considerations for intraoral use (Diode lasers) 2
Historical background and recent advances (Diode lasers) 3
Regulatory status and current clinical uses (Diode lasers) 4
Product trends and future developments (Diode lasers) 4

Section 8: Erbium Lasers

Technical background—Erbium lasers 1
Technical background—tissue interactions and effects (Erbium lasers) 2
Technical background—considerations for intraoral use (Erbium lasers) 3
Optical fibers for erbium laser delivery (Erbium lasers) 4
Historical background (Erbium lasers) 4
Recent advances (Erbium lasers) 5
Regulatory status and current clinical uses (Erbium lasers) 5
Product trends and future developments (Erbium lasers) 5

Section 9: Holmium Lasers

Technical background—Holmium lasers 1
Technical background—tissue interactions and effects (Holmium lasers) 2

Technical background—considerations for intraoral use (Holmium lasers) 4
Historical background and recent advances (Holmium lasers) 5

Section 10: Light-Emitting Diodes (LEDs)

Technical background—LEDs 1
Technical background—material interactions and effects (LEDs) 2
Technical background—considerations for intraoral use (LEDs) 3
Historical background and recent advances (LEDs) 3
Regulatory status and current clinical uses (LEDs) 3
Product trends and future developments (LEDs) 3

Section 11: Free-Running, Pulsed Nd:YAG Lasers (1064 nm)

Technical background—Pulsed Nd:YAG lasers 1
Technical background—tissue interactions and effects (Pulsed Nd:YAG lasers) 2
Technical background—considerations for intraoral use (Pulsed Nd:YAG lasers) 3
Historical background and recent advances (Pulsed Nd:YAG lasers) 4
Regulatory status and current clinical uses (Pulsed Nd:YAG lasers) 4
Product trends and future developments (Pulsed Nd:YAG lasers) 5

Section 12: Lasers vs. Electrosurgical Instruments

Laser modalities 1
Electrosurgical modalities 1
Performance issues 2
Surgical precision 2
Cutting, coagulation, and vaporization 3
Compatibility with irrigation methods 6
Neuromuscular stimulation 7
Tactile feedback 7
Compatibility with endoscopic instrumentation and techniques 8
Safety issues 9
Ergonomic issues 11
Cost issues 12
Reimbursement issues 13

Medical Lasers

This section defines terminology used throughout the book, and attempts to provide the non-laser-technical reader with the basic technical background needed to better understand new developments in medical laser products and technology. It is not intended to be a comprehensive treatment of the subject, but instead is written with the specific needs of this book in mind.

Laser energy and biomedical work

Light bulbs and lasers both generate light, which is the common name for electromagnetic energy we can see. However, as will be discussed in more detail below, the special characteristics of laser light render it much more useful for medical and other applications.

Energy is the ability to do work. Light or electromagnetic energy is but one form of energy, with other forms being heat, chemical energy, nuclear energy, mechanical energy, and electrical energy. Examples of biomedical work that can be done with laser energy include: raising the temperature of biological tissue until it is cooked or vaporized, mechanically breaking apart kidney stones, and causing specific biochemical reactions to occur in tissue.

A frequently used unit of energy is the ***Joule (J)***. To give an idea of how much work can be done with 1 Joule of energy, consider that about 2500 Joules are required to heat and convert 1 cubic centimeter of water (about the size of a sugar cube) at body temperature (98.6°F) into steam at 212°F; that is, to completely boil it away as steam. This example is particularly relevant to laser surgery applications, where it is often the objective to remove biological tissue, which is mostly water, by converting it into steam or vapor. Converting tissue into vapor, or *vaporizing* it, allows tissue to be removed by vacuuming it away with a handheld suction instrument.

Another unit of energy commonly used in laser parlance is the *milliJoule (mJ)*, which is one thousandth of a Joule. It is used when low energies are being discussed, so as to avoid the use of decimals. For example, a laser energy of 0.540 J can also be expressed as 540 mJ.

Power is the rate of doing work. Lasers that generate more power, or more energy per unit of time, are capable of doing more work per unit time than lower power lasers. Conversely, higher power lasers can do the same amount of work in less time than less powerful lasers. In our water cube example, a more powerful laser would completely boil or vaporize the cube faster than a less powerful laser. Various units for expressing laser output power are used depending on the power capability of the laser being described. Most lasers used in laser surgery generate power levels of 1 to 100 watts, where one *Watt (W)* is equal to an energy generation rate of 1 Joule per second. Assuming that all of the laser light shined on our water cube could be used to raise its temperature, a 1-watt laser would vaporize the water cube in 2500 seconds (40 minutes), whereas a 100-watt laser would vaporize it in 25 seconds.

Some lasers, such as those used in laser lithotripsy applications (fragmentation of kidney, ureter, or gall stones), generate power levels as high as 100,000 watts. Such a high power level would typically be referred to as 100 KiloWatts (100 KW), where 1 *kilowatt* = 1000 watts. Lasers used for doing cataract-related eye surgery generate *MegaWatt* (MW) power levels, or levels as high 1,000,000 watts (1 MW). Finally, lasers used for welding detached retinas employ power levels in the *milliWatt* (mW) range, where 1 milliWatt = one thousandth of a Watt (.001 watts).

Lasers as energy and power conversion devices

As they are used in therapeutic medical applications, lasers are, in essence, devices that convert electrical power provided at a wall outlet into a monochromatic and directional beam of light powerful enough to do biomedical work. In most cases, it is only when light is provided in such a form that it can be used to create medically useful tissue effects. The terms "monochromatic" and "directional" are defined below in the context of a light-bulb analogy.

Light bulbs convert electrical energy into light energy; heat energy is also generated as an unwanted by-product (light bulbs are hot to touch). For example, a 100-watt light bulb consumes 100 watts of electrical power and converts this into about 20 watts of white light energy that can be detected by the human eye (we're guessing as to what the actual amount is). The other 80 watts of electrical power goes into raising the temperature of the light bulb's glass envelope and other nearby structures, and into light "colors" that cannot be detected by the naked eye (sometimes referred to as *radiant heat*—your hand gets warm if you hold it near, but not in contact with, the light bulb). Similarly, lasers convert electrical energy into light energy generating some waste heat in the process.

The light bulb generates "white" light, or light of all different colors or **wavelengths**, and emits this light in all directions. One could convert a light bulb into a reasonably **directional** light source (a flashlight basically) by completely enclosing the bulb in a box and putting a small hole in one side of the box. The light beam emerging from the hole would spread with distance, or **diverge**, so that light rays would be emitted into a limited range of different directions, but they would all be headed in the same general direction. Note, however, that most of the light energy generated by the light bulb hits the inside walls of the box where light is absorbed and converted into useless heat. Only a small proportion of the light is emitted through the hole as a directional light beam.

One could then convert the light bulb into a reasonably directional and **monochromatic** (one-color) light source by placing a colored filter over the hole in the box. This filter would allow a narrow range of light wavelengths to emerge that might be red in color, for example. The filter would absorb light energy at all other wavelengths and convert it into heat (the filter would heat up in the process). Now we have a flashlight that emits a red light beam. Note that, in the process of filtering out all light except the red wavelengths, we have further reduced the power in the emitted beam. We would be doing well to have a few microwatts (millionths of a watt) of red light power at this point, which isn't enough to be of any use for most therapeutic medical applications.

Although it is an oversimplification to say, most medical lasers aren't much different, in principle, than the filtered-lamp light source just described considering how lasers are currently used for therapeutic applications. Lasers differ primarily in *the degree* of monochromaticity and directionality provided in that the range of light wavelengths and the range of directions into which they are emitted, are much smaller than what can be achieved easily with a light-bulb source. More importantly, lasers are able to convert electrical power into a monochromatic and directional light beam *much more efficiently;* that is, with much less electrical energy converted into waste heat. Instead of a few microwatts of power, a small laser that consumes 100 watts of electrical power might convert this into 10 watts of directional and monochromatic light, or almost 10 million times more power than what was achieved in the light-bulb analogy. In this form, 10 watts of light power is more than enough to be useful for many therapeutic laser applications. Throughout this report we refer to *laser light,* or **laser beams,** to indicate we are using light energy that is highly monochromatic and highly directional.

Basic components of a laser

In general, lasers consist of five basic components or subsystems: the active medium, the laser resonator, the power supply, the cooling subsystem, and the controller. The **active medium** is a material of some sort (solid, liquid, or gas) where laser light is generated via a process called **stimulated emission**. Stimulated emission occurs automatically when certain design and operating conditions are met, as described below. For the most part, the active medium used to construct the laser determines the output wavelength(s), power, and energy generated by the laser, and is chosen according to the needs of the particular medical application at hand. Different types of lasers are typically referred to by the name of the active medium employed, *e.g.*, carbon dioxide lasers, Nd:YAG lasers, holmium lasers, etc.

The active medium is physically located or positioned within an optical subsystem called the laser **resonator.** In its simplest form, the laser resonator consists of two mirrors separated by some distance (12" for example) and the mirrors are *aligned* so their reflecting surfaces face each other and are parallel. Light traveling in a direction perpendicular to the mirror surfaces will bounce back and forth very many times between the mirrors. On the other hand, light traveling in a different direction will either not hit the mirrors at all, or will make only a few passes between the mirrors before it misses one of the mirrors and is effectively removed from the laser resonator.

The active medium is positioned between the mirrors. The active medium generates light at the wavelength of interest in response to being excited, or *pumped*, by the laser's *power supply*. Some of this light is emitted in the direction perpendicular to the two parallel mirror surfaces, and bounces back and forth between the mirrors, making very many passes through the active medium. Each time the light passes through the active medium its power is increased or *amplified*. Only after the beam has made very many passes through the active medium is its power high enough to be useful. One of the resonator mirrors reflects only a portion of the light incident on its surface, 80% for example, letting the other 20% "leak out" one end of the resonator. This optically leaky mirror is called the *output coupler*. Laser light emerges through the output coupler as a monochromatic and directional beam of energy.

Not all of the power put into the active medium by the laser's power supply is converted into laser light. Some of the input power is converted into heat and raises the temperature of the active medium. This heat must be removed from the active medium, and the active medium's temperature must be maintained at or below some maximum operating temperature. This is accomplished by the laser's *cooling subsystem* or *cooler*.

Finally, there is the laser's *controller* subsystem, which, in most medical lasers, is a microcomputer or microprocessor located somewhere inside the laser. The laser has a *control panel* that functions as a computer input device (similar to a computer keyboard) and is used by the operator to tell the controller what the desired laser power and other output parameters are (see below). The controller sends appropriate electronic signals to the laser's power supply, resonator, and cooling subsystems so the desired laser emission is produced.

Solid-state, liquid dye, and gas lasers

There are three general types of lasers that will be discussed. These are categorized based on the physical state of the active medium used to make the laser: solid-state lasers, gas lasers, and liquid dye lasers. In a **solid-state** laser, the active medium usually consists of a crystalline material fabricated into the shape of a rod about the size of cigarette. This **laser rod** typically has ends that are polished very flat and smooth so laser light can pass back and forth through the length of the rod without optical distortion or scattering. Examples of solid-state active media and lasers include neodymium:YAG (Nd:YAG), holmium:YAG, and alexandrite. **Semiconductor diode lasers** are a special kind of solid-state laser and are usually referred to separately as such. Solid-state laser rods are usually excited into stimulated emission and lasing, or *pumped*, optically, using white light from a flashlamp or arc lamp, or the monochromatic light emission from another laser.

In a **gas laser**, the active medium generally consists of a hollow tube with windows on the end and filled with an appropriate laser gas or mixture of gases. A **carbon dioxide** (CO_2) laser uses a mixture of carbon dioxide, nitrogen, and helium gas, but only the carbon dioxide gas molecules undergo stimulated emission to generate the CO_2 laser wavelength used in laser medicine (the other gases are included to improve laser output power). **Excimer** lasers are gas lasers that employ mixtures of argon, krypton, or xenon gas with fluorine or chlorine gas. Other types of gas lasers used in laser medicine include **argon ion** (argon) lasers, **krypton ion** (krypton) lasers, and **metal vapor** lasers. To pump the active gas or vapor medium, an electrical current is passed through the gas mixture to create an **electrical discharge** in the medium in much the same way that electrical discharges in the glass tubes of neon signs are created.

A solution consisting of an organic dye, similar to the dyes used to color clothing, dissolved in a methanol- or water-based solvent forms the active medium for a **liquid dye laser**. The dye solution flows through a cylindrical dye cell with windows on its ends, or is literally squirted through the laser resonator in the form of a dye stream or jet. Flashlamps or other lasers are used to optically pump dye lasers. Liquid dye lasers used in laser medicine include **flashlamp-pumped dye lasers** and **532 nm laser-pumped dye lasers**.

Although there are exceptions, solid-state medical laser products tend to be more compact and portable than gas or liquid dye lasers of the same power capability. Furthermore, the active medium of a solid-state laser is not consumed by the laser process. Handling of potentially hazardous materials, and operating costs associated with periodically replenishing laser liquids or gases, are eliminated by use of a solid-state design. Such ergonomic and economic advantages usually favor the use of a solid-state laser, but only if one is available that provides the wavelength, power, and other laser emission parameters needed to perform the intended therapeutic procedure.

At this time, whether a gas, liquid dye, or solid-state laser is used is still determined primarily by the requirements of the application at hand in terms of specific combinations of laser output parameters. However, the inherent laser output versatility afforded by solid-state design techniques is increasingly resulting in new solid-state laser products that are more versatile medical instruments than their gas or dye laser counterparts. In other words, the number of therapeutic laser procedures that can't be performed with a solid-state laser appears to be decreasing.

Optical pumping

As we mentioned earlier, a laser's active medium generates laser energy in response to being excited or pumped by the laser's power supply. However, except for the case of semiconductor diode lasers, electrical power is not used directly to excite the active medium into lasing. Instead, the laser rod of a solid-state laser, or the dye cell of a liquid dye laser, is pumped with light energy. This is referred to as **optical pumping**. As we have defined the power supply, some part of it generates excitation or pump light that is somehow coupled or shined into the laser rod or dye cell. The power supply converts electrical power from the wall outlet into pump light that has characteristics needed to successfully excite the laser rod into lasing.

Optical pumping sources include flashlamps, arc lamps, and other lasers. Flashlamp and arc lamps are really just special types of light bulbs designed for use in lasers. They emit white light in all directions, all or most of which must be collected and shined into the laser rod or dye cell if the laser is to operate efficiently. This is usually accomplished by placing the lamp, which is also rod-shaped and about the size of a cigarette, physically next to, and parallel with, the laser rod (or dye cell). The laser rod and lamp are enclosed in a ***pump cavity*** with internal walls that are highly polished and designed to reflect (direct) the white light emitted by the lamp into the laser rod. ***Flashlamps*** are used in pulsed lasers (see below) and operate much like a camera flashbulb. ***Arc lamps*** are used in continuous-wave solid-state lasers and are similar to the very bright lamps used in searchlights or cinema projectors.

As we have just described them, lamp-pumped solid-state and liquid dye lasers are devices for converting white-light emission from a flashlamp or arc lamp into monochromatic and directional light beams. The process of stimulated emission, as it occurs in a laser resonator, provides a much more efficient means for doing this than putting a light bulb in box, punching a hole, and putting a colored filter over the hole. As a result, much higher power beams are generated that are useful for therapeutic medical applications.

A special case of optical pumping is ***laser-pumping***. The output from another laser is used to pump some types of solid-state lasers and liquid dye lasers. The point of doing so is usually to obtain some new laser output performance capability, such as a different wavelength, that is not available from the pump laser itself. In some instances, lamp pumping may not be possible at all, requiring the use of a laser pump source to excite the desired active medium into lasing. A special case of laser-pumping is semiconductor diode-laser-pumping, or ***diode-pumping***, which, when it can be used, offers many important advantages over lamp pumping and pumping with other lasers.

Solid-state laser media

The active medium of a solid-state medical laser is typically a rod about the size of a cigarette. Rod ends are usually polished flat and very smooth so laser light can pass through the laser rod without optical distortion as it bounces back and forth between the resonator mirrors. Rod diameters of 3 to 6 millimeters and lengths of 2" to 4" (50 to 100 mm) are common in lamp-pumped laser systems. Actual rod dimensions vary depending on the specific laser product design (output power, energy, etc.). Laser rod sizes are generally much smaller in diode-pumped solid-state lasers (see below).

Solid-state laser rods consist of a *host crystal* material into which a small amount of *dopant* atoms are uniformly distributed. This is accomplished when the host crystal is being grown by adding a small amount of dopant material to the "melt" from which the crystal is "pulled". Examples of atoms commonly used as dopants in solid-state laser rods include neodymium (Nd), chromium (Cr), erbium (Er), holmium (Ho), and thulium (Tm). The letter designations shown in parentheses are the *chemical symbols* for each atom and are the same symbols one would see on a periodic chart of elements hanging in a high-school chemistry class. As they exist in the laser rod, these atoms have one, two, or three of their electrons removed, so that, strictly speaking, they are *ions* rather than atoms. However, we use the terms "ion" and "atom" interchangeably.

In general, dopant atoms, rather than host-crystal atoms, are responsible for emitting laser energy via stimulated emission. The host material is there basically to hold the dopant atoms in place and to provide a physical environment for dopant atoms that allows them to participate in the stimulated emission (lasing) process. The host material must be transparent at light wavelengths used for optically pumping the dopant atoms so that pump light can reach all dopant atoms, including those at the center of the laser rod. Pump light is absorbed by the dopant atoms, exciting them into an energy state that allows them to undergo stimulated emission. The host crystal must also be transparent at the laser wavelength so laser light can pass back and forth through the rod and be amplified by the stimulated emission process.

The most commonly used laser host crystal is *yttrium aluminum garnet* or *"YAG"* for short (yttrium is pronounced "it-tree-um"). This material consists of yttrium, aluminum, and oxygen ions arranged in a specific three-dimensional pattern that gives the YAG crystal physical properties that are very useful for laser applications. Highly transparent and distortion-free laser rods can be fabricated out of YAG crystals and the rods are economical to make and purchase. YAG is mechanically durable and doesn't scratch or break easily. The YAG material has good thermal conductivity, which means it is fairly easy from a design standpoint to remove heat generated in the rod during the lasing process (that is, to cool the laser rod without compromising other laser design requirements). Examples of other types of host crystals used in solid-state lasers include YLF (yttrium lithium fluoride), sapphire (similar to the gem stone, but colorless), chrysoberyl (similar to sapphire), and YSGG (see below). These alternative host materials are often used where YAG can't be used to make the desired laser, or because they provide some design advantage over YAG that is important for the intended application. Laser rods fabricated out of these other host crystals are usually more expensive than YAG by a factor of two or more.

The *symbology* used to designate solid-state laser materials consists of the symbol of the dopant ion separated from the host crystal designation by a colon. For example, neodymium-doped YAG is designated Nd:YAG. Similarly, chromium-doped YSGG is written Cr:YSGG. Sometimes more than one dopant atom is used in a laser rod, in which case dopant atom symbols are separated by commas. YAG doped with holmium, chromium, and thulium atoms is designated as Tm,Ho,Cr:YAG.

Chromium-sensitized laser host crystals improve the laser efficiency of some types of flashlamp-pumped solid-state lasers (allow more laser energy to be generated for the same amount of electrical input energy). These are host crystals doped with chromium ions in addition to the dopant ions that generate laser energy at the desired wavelength (the laser ions). Chromium ions are included because they are capable of absorbing more of the light wavelengths emitted by a (white-light) lamp pumping source, compared to the laser ions typically used to dope solid-state laser crystals (Nd, Ho, Tm, Er). Once a chromium ion has been pumped into an excited energy state, it can then *transfer* its energy to a nearby laser ion if an appropriate host laser crystal is used and other composition conditions are met. In host crystals compatible with using them, chromium ions allow a higher fraction of pump-light energy to pump the intended laser ions, thereby *sensitizing* the laser ions to pump wavelengths that would normally be wasted. More pump light energy is converted into laser emission, rather than wasted as heat, and laser output efficiency is improved. The chromium ions might also undergo stimulated emission, or lase, if allowed to do so, but the laser resonator is typically designed to prevent this.

In most cases, chromium ions cannot act as an effective sensitizer in a YAG crystal for holmium (Ho), erbium (Er), or neodymium (Nd) ions; that is, the chromium ions cannot directly transfer their energy to these other ions when in a YAG host crystal. This is unfortunate in the case of an Nd:YAG laser considering the many medical uses these lasers have. Some other host crystal must be used instead. In most cases, gallium garnet crystals, such as GSGG (gadolinium scandium gallium garnet) or YSGG (yttrium scandium gallium garnet), are used to obtain the benefits of chromium-sensitization. Lamp-pumped lasers that use the neodymium-doped gadolinium material (Nd,Cr:GSGG) are about twice as efficient as the equivalent Nd:YAG laser.

Perhaps the best example of where chromium-sensitized laser materials have made an impact in laser surgery is in the design of holmium lasers used in arthroscopic surgery, general surgery, ophthalmology, and cardiovascular surgery. These holmium lasers use a thulium- and holmium-doped chromium-sensitized YAG crystal, designated as Tm,Ho,Cr:YAG or THC:YAG. This material represents one of the few situations where chromium-sensitization does work well in a YAG host crystal. Ho:YAG lasers (no thulium or chromium ions) have been used in the laboratory for more than a decade but never appeared as commercial medical devices. The Ho:YAG laser rod must be cooled with liquid nitrogen to get reasonable output efficiency, but even then efficiency is not as good as one would like for a medical product. Furthermore, most would agree that lasers with (flowing) liquid nitrogen cooling systems are not convenient in a clinical setting. The innovation that THC:YAG made possible is that it enables good holmium laser efficiency *at room temperature.*

It is not just chromium-sensitization that allows THC:YAG to work so efficiently. The presence of thulium ions is absolutely required considering that chromium ions cannot directly transfer pump energy to the holmium ions. Thulium ions are able to accept energy from excited chromium ions and efficiently transfer this energy to holmium laser ions. New understanding of the multi-step energy transfer processes that can occur in solid-state laser materials will likely result in new lasers and other improved lasers.

For most of their history, which began in 1960, solid-state lasers have employed cylindrical or rod-shaped laser rods. Cylindrical rods are relatively easy to fabricate and are easy to design into lamp-pumped laser systems (to cool effectively, for example). However, rod-shaped geometries have certain inherent limitations that laser designers would like to circumvent. One of the more important limitations, as far as medical applications are concerned, has to do with how much laser power can be delivered efficiently through an optical fiber delivery system (see next section). Due to a process called **thermal lensing** that occurs in rod-shaped laser crystals, the amount of laser power that can be coupled into and delivered by an optical fiber often limits the amount of power that can be used for minimally-invasive treatment procedures. For some lasers, it may be possible to fiberoptically deliver only a fraction of the power the laser is capable of generating. The problem of thermal lensing tends to be particularly severe in chromium-sensitized laser materials.

Slab laser crystal geometries are being investigated that greatly reduce thermal lensing. The laser crystal is fabricated in the shape of a small rectangular bar or "slab". Thermal lensing is reduced partially as a consequence of the fact that the active laser medium's cross-section is rectangular rather than circular. Slab laser technology is fairly new, and laser designers are still learning how to best design and operate ***slab lasers***. Slab laser crystals are considerably more expensive to fabricate than laser rods, and a pump cavity that can accommodate a slab crystal is also more complex and expensive. Nevertheless, the increased costs may be warranted for some applications if slab designs allow significantly higher laser power to be delivered through a small-diameter optical fiber.

Laser wavelengths and wavelength versatility

We have alluded to the color of laser light as its wavelength, but the term is more broadly applicable than to what can be seen as colored light. To the extent that laser light can be thought of as periodic *waves* of energy traveling through space, **wavelength** refers to the physical distance between crests of successive waves in the laser beam. However, instead of inches, feet, or meters, laser wavelengths are measured in units of length called nanometers or microns. By definition, 1 **nanometer (nm)** = 10^{-9} meters, or one-billionth of a meter (a meter is 39 inches). One **micron (μm)** is equal to 10^{-6} meters, or 1000 nm. Typical medical laser wavelengths include 193 nm,

308 nm, 630 nm, 800 nm, 1064 nm (1.06 microns), 2100 nm (2.1 microns), and 10,600 nm (10.6 microns).

Only laser wavelengths between 400 nm and 700 nm are visible to the unaided eye and are associated with named "colors." For example, colors associated with some visible medical laser wavelengths are: 488 nm (blue), 532 nm (green), 577 nm (yellow), 590 nm (orange), and 694 nm (deep red). Other wavelengths near the ones listed would have similar colors; the transition between colors is gradual as wavelength changes.

There are, however, other medically useful laser wavelengths the eye cannot see and that are not associated with colors. Wavelengths longer (numerically greater) than 700 nm are referred to as *infrared or IR* wavelengths, whereas wavelengths shorter (numerically less) than 400 nm are called *ultraviolet or UV* wavelengths. Infrared wavelengths are often further categorized as *near-IR* (700-1200 nm), *mid-IR* (1200-3000 nm), and *far-infrared* (FIR; 3000 nm and greater), although the demarcation points tend to vary considerably among different authors. Ultraviolet wavelengths are often referred to as *near-UV or UVA* (400-320 nm), *mid-UV or UVB* (320-290 nm), and *far-UV or UVC* (290 to about 200 nm).

Laser wavelength is an important factor in most types of laser-tissue interactions and often dictates the kinds of therapeutic procedures that can be performed easily with a given laser. Accordingly, an important trend among newer medical products is to provide some amount of *wavelength versatility* that allows the operator to select among two or more different wavelengths at the laser control console. For example, newer frequency-doubled Nd:YAG surgical lasers may allow one to switch quickly between 532 nm and 1064 nm. The 532 nm wavelength is used for precise cutting, vaporization, or superficial tissue coagulation, whereas 1064 nm may be used for deep coagulation of tissue. The operator simply pushes a button and, within a matter of seconds, output wavelength is switched.

Tunable lasers represent an extreme case of wavelength versatility. These are lasers whose output wavelength can be varied continuously throughout some range simply by adjusting an appropriate control knob. The ability to "tune" the laser output wavelength is basically a characteristic of the active laser medium itself. Examples of tunable solid-state lasers that could appear as medical devices include alexandrite, cobalt magnesium fluoride, and titanium sapphire lasers. Liquid dye lasers are also continuously tunable lasers.

Nonlinear wavelength conversion

Various **nonlinear optical processes** can be used to convert laser energy at one wavelength into laser energy at another wavelength. Nonlinear wavelength conversion is one way to add wavelength versatility to a laser. Wavelength conversion is usually accomplished by passing a laser beam through a suitable nonlinear optical device that might employ a gaseous, liquid, or solid-state nonlinear optical material of some sort. In passing through the nonlinear device, a portion of the input laser beam's energy is **converted** and one or more new laser beams with different wavelengths emerge. The conversion process is typically not 100% efficient, so that a laser beam with the original wavelength emerges as well. Nonlinear optical components are often passive devices that do not require a power supply of any sort, which implies that the total laser power of all wavelengths emerging from the device can be no greater than the input laser power.

Second harmonic generation, or **frequency-doubling**, is the nonlinear process by which the input laser wavelength is divided by a factor of two (laser frequency and wavelength are inversely related—doubling the frequency cuts the wavelength in half). The frequency-doubled Nd:YAG lasers used in laser medicine have a **fundamental** wavelength of 1064 nm, which is an invisible infrared wavelength. When the beam is sent through a second harmonic generation (SHG) crystal, a green laser beam emerges with a wavelength of 532 nm (1064 ÷ 2). A doubling efficiency of 50% is typical, which means that about 5 watts of green output is produced from 10 watts of input power at 1064 nm. The other 5 watts emerges from the crystal as unconverted 1064 nm power. The two beams are separated optically so that only one wavelength is used at a time.

The nonlinear crystal used most often in frequency-doubled medical lasers, which at this time are almost exclusively Nd:YAG lasers, is a material known as potassium titanyl phosphate or KTP ("K" is the chemical symbol for potassium). KTP is a solid-state crystalline material that was invented and patented by DuPont. As a result, frequency-doubled Nd:YAG medical lasers are often referred to generically as "KTP lasers". However, frequency-doubled Nd:YAG surgical products are now available that do not use KTP, so it is probably more appropriate to refer to this class of lasers as frequency-doubled Nd:YAG lasers.

Third harmonic generation (THG) processes reduce the input laser wavelength by a factor of 3. In the case of an Nd:YAG laser, an ultraviolet wavelength of 355 nm is produced from a 1064 nm input beam. This is accomplished in two steps in most Nd:YAG laser designs. The 1064 nm beam is frequency-doubled in a first nonlinear crystal. The 532 nm and 1064 nm beams that emerge are then sent through a second *sum-frequency mixing* crystal thereby generating the 355 nm wavelength. This detail is not important here except to note that what we call frequency-tripled lasers are really "sum-frequency-mixed" lasers.

A **Raman shifter** is a nonlinear optical device used for shifting the input laser wavelength to longer or shorter wavelengths (SHG and THG processes always shift to a shorter wavelength). Furthermore, the amount by which the wavelength is shifted depends not only on input laser wavelength, but also on the actual nonlinear Raman medium that is used. Hydrogen gas will shift a 1064 nm input beam by a different amount than methane gas or benzene liquid. One must select an appropriate combination of input laser wavelength and Raman medium to get the desired output wavelength. Lasers that generate a wavelength of 1900 nm by Raman-shifting the 1064 nm output of an Nd:YAG laser have been investigated.

As in the case of laser crystals, the development of new and improved nonlinear optical materials is expected to increase the range of wavelengths available from solid-state laser devices and to result in more efficient and more powerful lasers at currently available wavelengths.

Laser beams

When laser light emerges from a laser, it usually does so in the form of a pencil-thin **beam** of laser energy traveling at the speed of light (186,000 miles per second). This beam travels in a straight line until it hits something that *reflects* it (changes its direction) or until it hits something that stops the beam (absorbs beam energy and converts into heat). For the purposes of this report, laser energy appears instantaneously at the point where the beam is being stopped, which is usually where it is being used to treat tissue, when the laser is activated.

Most lasers are designed to generate a beam with a circular cross-section. Depending on the design of the medical laser, the diameter of the beam at the laser output port is typically in the range of 3 to 6 millimeters (6 mm is about 1/4 inch). The laser beam *diverges* as it travels away from the laser, which means that the beam's diameter increases with distance from the laser. If one were to place a card or other flat object in the laser beam so the beam is perpendicular to the card, one would see a circular *laser spot* on the card with a diameter that increases as the card is moved away from the laser. The amount that spot diameter increases over a distance of 1 meter, for example, is a measure of the beam's *divergence*.

When we say that a laser beam is highly directional, we mean that its beam divergence is very small (the more divergent a beam is, the more directions light energy travels in simultaneously). Light beams with very low divergence are important for medical applications because they can be focused to very small spot diameters (see below). However, laser beams that have low divergence, which are said to be *collimated*, are also useful because they can be delivered through an articulated-arm or similar delivery system without having the beam diameter increase to an unmanageable size (see Section 3).

There are many instances in laser medicine and surgery where one needs a different beam diameter than that that generated by the laser, and where one deliberately introduces positive or negative beam divergence so as to change beam diameter. One typically uses an optical *lens* to do so. Passing a well-collimated laser beam through a *negative or diverging lens* can drastically *increase* the divergence of the beam. The beam diameter might double or triple over a distance of only a few inches from the lens. On the other hand, a *positive or focusing lens* causes the beam diameter to *decrease* with distance, or converge, until it reaches a distance from the lens known as the lens *focal length* where the beam is reduced to some minimum diameter. The beam is said to be *focused* at the focal length distance, or at the focal point. The beam diverges for distances beyond the focal point. For example, a focusing lens might be used to decrease beam diameter from a few millimeters down to about 0.050 mm (50 microns) over a distance of 1 to 2 inches.

One reason why laser beam diameter is manipulated in laser medicine and surgery is to alter the beam's **power density**. Power density, also called **intensity**, is measured in units of **watts per cm^2** (watts per square centimeter). Like wavelength, power density frequently determines the kinds of laser-tissue interactions and tissue effects that can be induced and controlled with a given laser. Power density (in watts per square centimeter) is numerically equal to the laser's output power (in watts) divided by the area of the beam at tissue or other point of measurement (area expressed in square centimeters). When one uses a lens to focus a laser beam, the area of the beam at the focal point can easily be 10,000 times smaller than the area of the beam at the point where it enters the lens (beam *diameter* at the focus would be 100 times smaller than at the lens, since area is proportional to diameter squared). Since a good lens does not significantly alter the amount of laser power in the beam, power density can therefore be 10,000 times higher at the focus than at the lens. Since beam diameter and area increase continuously on either side of the focal point, power density decreases continuously. A convenient way to control laser-induced tissue effects is to focus laser energy onto tissue with a lens and vary the distance between the lens and tissue surface so as to vary the beam's power density at the surface. One can also vary power density by varying the laser's output power, and maintaining a constant lens-to-tissue distance, but this is often less convenient than manipulating lens distance.

In addition to determining the kind of laser-tissue interaction that is induced, laser beam diameter is also typically adjusted to control the area of tissue irradiated or "treated" simultaneously. The diameter of the laser beam at the tissue surface is typically referred to as the ***irradiated spot size*** or ***treatment spot size***. Assuming sufficient power is available from the laser, one can simultaneously treat a larger area of tissue by using a larger spot size. (More laser power is needed to maintain the same power density as treatment spot size is increased). Treatment time is reduced because less repositioning of the beam is needed to irradiate the entire area to be treated. On the other hand, one uses very small laser spot diameters when doing procedures that require high precision; that is, when one wants to treat very small tissue structures without also irradiating nearby structures.

Temporal emission modes:
pulsed vs. continuous-wave (CW) lasers

Lasers emit light energy in either a **continuous-wave** or a **pulsed** temporal emission mode. Basically, these differ in how the emission of laser energy proceeds in time once the laser is activated.

An example of a continuous-wave (CW) light source is a light bulb. The bulb begins emitting light when the light switch is turned on, and continues to do so until the switch is turned off. Similarly, CW lasers begin emitting laser energy when they are first turned on and do so continuously until they are turned off. In the case of a medical laser, the surgeon usually starts and stops laser emission by pressing or releasing a footswitch (footpedal).

Examples of pulsed light sources include camera flashbulbs and strobe lights. The camera flash is a **single-pulse** source in that one pulse of light is emitted each time the camera button is pressed. A strobe light is a **repetitive-pulse** light source in that pulses are emitted periodically for as long as the firing button is pressed. The same terminology applies to pulsed lasers, which can typically be operated in a single-pulse or repetitive-pulse mode.

Note that one can mimic a pulsed light source with a light bulb by manually flicking its switch on and off quickly. We refer to a light bulb operated this way as a **quasi-CW** light source. A CW laser can also be operated in a quasi-CW mode, but the laser is usually "flicked" on and off automatically by the laser's controller rather than by the operator. The basic difference between quasi-CW and pulsed laser sources, as we use these terms, is that a pulsed laser emits much shorter duration light pulses than can be generated by a quasi-CW laser. The short pulses typical of pulsed lasers allow them to emit much more powerful light pulses than a quasi-CW laser. The terminology can be confusing, since quasi-CW lasers and pulsed lasers are often described as operating in a single-"pulse" or repetitive-"pulse" mode.

For the purposes of this book, medical laser products described generically as ***pulsed lasers*** are configured to operate only as pulsed lasers and cannot be operated by the user in a CW emission mode. On the other hand, CW laser products can almost always be operated in one or more quasi-CW emission modes by the user, but cannot be operated in a short-pulse-duration, high-peak-power emission mode (see below for what is meant by short pulse duration and high peak power).

Unfortunately, the jargon is more extensive than just CW, quasi-CW, or pulsed laser emission modes. One also refers to several different types of quasi-CW operation, such as single-pulse, chopped, and super-pulsed emissions. Different types of pulsed modes include free-running, Q-switched, and mode-locked. Before attempting to describe these, we first define some of the basic terminology associated with pulsed lasers.

Pulse duration, pulse rate, peak power, average power

Pulse duration and ***pulsewidth*** are synonymous terms that refer to the length of time a laser actually emits during an individual laser ***pulse*** (like the brief duration of a camera flash). Depending on the type of laser and the pulsed emission mode in which it operates, laser pulse durations are measured in units of seconds, milliseconds, microseconds, nanoseconds, or picoseconds. One ***millisecond (ms)*** = 10^{-3} seconds = one thousandth of a second. One ***microsecond (µs)*** = 10^{-6} seconds = one millionth of a second. One ***nanosecond (ns)*** = 10^{-9} seconds = one billionth of second. One ***picosecond (ps)*** = 10^{-12} seconds = one thousandth of a nanosecond. One ***femtosecond (fs)*** = 10^{-15} seconds = one thousandth of a picosecond.

As was discussed earlier, laser *power* refers to the *rate* at which energy is generated by the laser. A laser power of 1 Watt means that 1 Joule of energy is generated in 1 second, or 0.5 Joules is generated in 0.5 seconds, or 0.1 Joules in 0.1 seconds, etc. The actual amount of laser energy generated depends on the length of time that the laser is left on at the 1 Watt power level, *i.e.*, on the pulse duration. A laser on for 0.1 seconds at a power level of 100 watts would generate 100 watts x 0.1 seconds = 10 Joules of energy, and would generate 20 Joules if left on for 0.2 seconds. Therefore, for a pulsed laser, the *pulse energy* generated in a single laser pulse depends on the pulse duration and the output power level during the pulse.

Peak power refers to the power level during an individual laser pulse. As described in the previous paragraph, pulse energy (in Joules) = peak power (in watts) x pulse duration (in seconds). Peak power, along with peak power density at tissue, often determine the **kinds of tissue effects** the laser can be used to induce in tissue (coagulation, vaporization, fragmentation, etc.)

In medical laser jargon, one refers to the pulse energy being used more often than one refers to peak power. Peak power can be calculated, however, assuming one knows the pulse duration and by rearranging the earlier expression. The (nominal) peak power of a laser pulse, expressed in watts, is numerically equal to pulse energy, expressed in Joules, divided by pulse duration (in seconds).

$$\text{(Nominal) peak power (watts)} = \text{pulse energy (Joules)} \div \text{pulse duration (seconds)}$$

For example, a laser that generates pulses each with 1 Joule of energy, and a 1 microsecond (10^{-6}) pulse duration, is said to generate a peak power of 1 Joule ÷ 10^{-6} seconds = 10^6 watts (1 million watts, or 1 megaWatt). The word "nominal" is used to indicate that this expression gives only an approximate value for peak power since the actual peak power value depends on the details of the laser pulse "shape" in time (triangular, rectangular, etc.).

In the case of a pulsed or quasi-CW laser, it is important to distinguish between "peak" and "average" output power. One usually cites average power values only in reference to lasers that are operated in a ***repetitive-pulse*** mode. Laser pulses are emitted periodically at some ***pulse rate***, such as 10 pulses per second (pps), 1 pps, or 1000 pps, so that pulses are spaced at regular intervals in time. A commonly used shorthand for pulses per second is Hertz (Hz) where 1 Hz = 1 pps. **Pulse period** is the separation in time between (the beginning of) successive pulses in the repetitive pulse train. The same units are used to express pulse periods as are used to express pulse durations. Pulse period is numerically equal to 1 ÷ (pulse rate) and is equal to 1 second for a laser operated at a 1 Hz pulse rate, and 0.1 seconds (100 milliseconds) for a laser operated at a 10 Hz rate. Similarly, a laser operated at a 0.5 Hz pulse rate generates one pulse every 2 seconds (pulse period of 2 seconds).

Unlike peak power, which is a measure of how much energy is generated during the brief duration of a single laser pulse, a laser's ***average power*** indicates the amount of energy generated *on a sustained basis*, over a period of seconds, for example, regardless of pulse duration. Average power is equal to the total energy measured, expressed in Joules, divided by the time in seconds over which the measured amount of energy is accumulated. Alternatively, average power is equal to pulse energy (in Joules) multiplied by pulse rate (Hz).

Average power (watts) = pulse energy (Joules) x pulse rate (Hz)

For example, a laser that generates 1 Joule per pulse at a 10 Hz pulse rate generates an average power of 10 watts (1 Joule x 10 Hz = 10 watts). If instead a 1-Joule pulse is generated every 2 seconds (a pulse rate of 0.5 Hz) then the average power is 1 Joule x 0.5 Hz = 0.5 watts = 500 milliwatts.

The average power of a pulsed or quasi-CW laser emission is, by definition, always less than the peak power of that emission. For quasi-CW emissions, average power can be comparable to peak power. In contrast, for "pulsed" laser emission, as we use the term, average power is usually one or more orders of magnitude (powers of ten) smaller than peak power.

Since there are no "pulses" to speak of, peak power is numerically equal to average power for CW laser emission. One cannot distinguish between peak and average power as defined above, and one refers only to "output power" or "CW output power." For CW laser emission, average power, CW power, and peak power all have the same numerical value.

Once laser ***peak power*** (or power density at tissue) has been selected to produce the desired ***type*** of tissue effect, a pulsed laser's ***average power*** is then adjusted according to ***how quickly*** the surgeon wishes to treat, or produce the desired tissue effect in, a given area or volume of tissue.

Medical Lasers

Quasi-CW operation: single pulse, chopped, super-pulsed

Most continuous-wave (CW) medical lasers can be operated in a single-pulse, chopped, or a superpulsed quasi-CW emission mode. Compared to CW operation, quasi-CW operation can provide the surgeon with enhanced surgical precision and control in some situations. Figure 1-1 shows how "true" CW emission proceeds with time. In this figure, and in Figures 1-2, 1-3, and 1-4, we assume the operator presses the footswitch at t=0 (time 0), and keeps pressing it forever. Laser emission begins virtually at the instant the laser's footswitch is pressed. Assuming the operator has selected a laser power setting of 100 watts, the output power level stays at 100 watts.

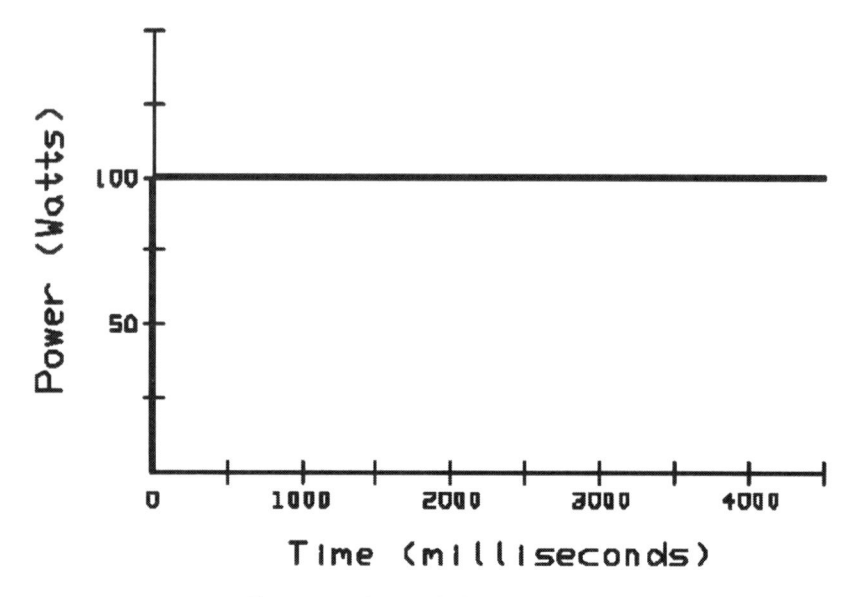

Figure 1-1 CW emission at 100 watts

Single-pulse quasi-CW emission is shown in Figure 1-2. In spite of the fact that the operator keeps pressing the footswitch, laser emission is halted automatically after a preselected pulse duration has elapsed, or 1.5 seconds in this example (1.5 seconds = 1500 milliseconds). Automatic termination of laser emission relieves the operator of this responsibility, which can result in improved surgical control or more consistent treatment of tissue. Many CW medical lasers provide a selectable range of pulse durations, or *laser ON* times as they are often called when talking about quasi-CW emissions, of 0.1 to 9.9 seconds.

Medical Lasers

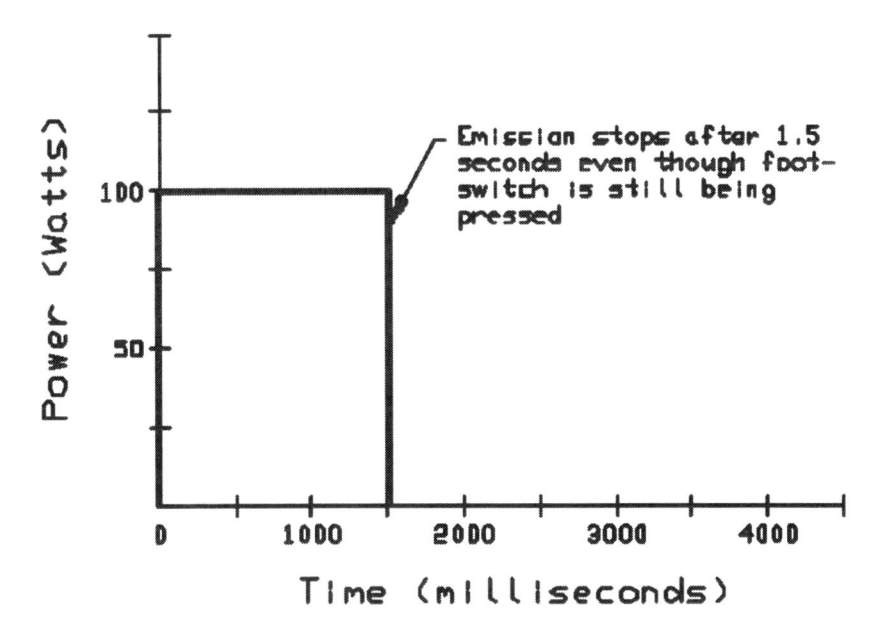

Figure 1-2 Singe-pulse quasi-CW emission

Chopped mode operation, also referred to as **repeat-pulse mode**, or **gated CW mode**, is shown in Figure 1-3. Pulses of a selected duration are emitted periodically at a selected pulse rate for as long as the footswitch is pressed. For most CW laser products, the operator selects a laser power setting (100 watts in this example), a laser ON time, and a *laser OFF* time. In the top diagram, laser ON time = laser OFF time = 500 milliseconds (0.5 seconds). The bottom diagram shows a laser ON time of 400 milliseconds with a laser OFF time of 100 milliseconds. Products generally provide a selectable range of OFF times similar or identical to the range of selectable ON times.

By definition, peak power in chopped emission mode is numerically equal to the CW power level setting that was selected by the operator (100 watts in this example). The surgeon might also have chosen a CW laser power setting something less than 100 watts, in which case the peak power in chopped mode would be less. The basic difference between chopped and CW emission is that chopped mode emission has periodic gaps of zero power level "chopped out of" the otherwise CW emission.

Medical Lasers

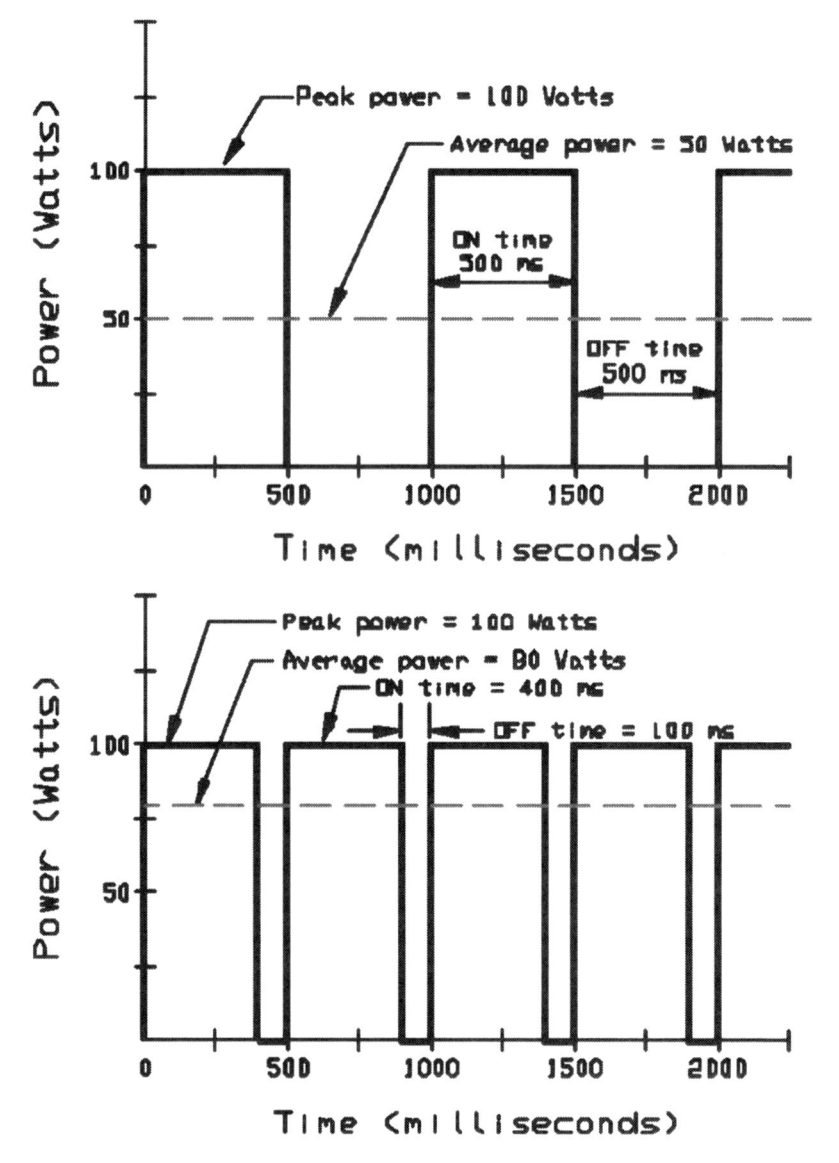

Figure 1-3 Chopped or gated cw emission

Figure 1-3 also shows how average power and peak power can differ. In the top diagram of Figure 1-3, pulse energy is 50 Joules (100 watts x 0.5 seconds). The pulse period in the top diagram is (0.50 sec + 0.50 sec) = 1 second, so pulse rate is 1 ÷ (1 sec) or 1 Hz. Therefore, average power is 50 Joules x 1 Hz = 50 watts, as shown by a dotted line on the graph, whereas peak power is 100 watts. In the lower diagram, pulse energy is 100 watts x 0.4 seconds = 40 Joules, and pulse rate is 2 Hz [1 ÷ (0.40 + 0.10 sec)]. Average power is 40 Joules x 2 Hz = 80 watts. Peak power is again 100 watts. Note that the average power output was increased from 50 to 80 watts simply by decreasing the OFF time between laser pulses, but peak power remained the same.

The top part of Figure 1-3 shows a repetitive pulse train that has a ***duty cycle*** of 50%. By definition, duty cycle is the ratio of laser ON time to the time between successive pulses or pulse period (ON + OFF time).

$$\text{Duty Cycle} = \text{ON time} \div (\text{ON} + \text{OFF time})$$

In the bottom diagram, the duty cycle is 400 ÷ (400 + 100) = 80%. Duty cycle is just another way of describing the OFF time between laser pulses. Note that a duty cycle of 100% (zero OFF time) means the emission is continuous in time, or CW. Note also that Duty Cycle x Peak Power = Average Power. In contrast to quasi-CW emission modes, where duty cycles of 10% to 100% are typical, "pulsed" medical lasers typically operate with a duty cycle much less than 10%.

Chopped mode operation is useful when the surgeon must quickly perform a repetitive laser treatment task over an area of tissue, such as coagulating an array of 1-mm-diameter spots over an area of several square inches. Laser ON time and power setting would be chosen to coagulate each spot with a single laser pulse and OFF time would be chosen to allow sufficient time for the surgeon to reposition the laser beam to the next treatment spot location. This would save the surgeon the effort of having to repeatedly press and release the footswitch, and might also reduce overall treatment time. Chopped mode operation can also improve the level of surgical control afforded to the surgeon. Since it normally takes a fraction of a second for the surgeon to actually release the footswitch, once the decision has been made to stop laser treatment chopped mode operation can reduce the amount of (unwanted) laser energy exposure that occurs during this reaction time.

Figure 1-4 shows ***super-pulsed*** operation in comparison with CW and chopped-mode operation. By definition, the maximum peak power in super-pulse mode is higher than the maximum power in CW mode by a factor that depends on the type of laser and the specific product design. For super-pulsed carbon dioxide (CO_2) lasers, peak power is typically 7 to 10 times higher than the maximum CW power provided by the same laser. The factor is more like 3 to 5 for CW Nd:YAG laser products. For example, a 100 Watt CW Nd:YAG laser may have a peak power of 300 watts in super-pulsed mode, as shown in the figure.

Also by definition, the duty cycle of a super-pulsed laser emission is always less than 100%. This contrasts with chopped or gated cw operation where duty cycle can typically be adjusted up to 100% if desired. The maximum selectable duty cycle is reduced (from 100%) by roughly the same factor that peak power is increased, so the maximum *average power* in super-pulsed mode is roughly equal to the maximum *CW power* level. In fact, for most super-pulsed products, maximum average power in super-pulsed mode is 10 to 20% less than maximum CW power.

In Figure 1-4, it is assumed that the maximum duty cycle of the product in question is 27%, so that the maximum average power is 80 watts (about 20% less than the CW maximum of 100 watts).

Superpulsing can reduce the amount of thermal necrosis and charring at cut tissue edges and otherwise improves surgical precision and control (see Section 2). Most carbon dioxide laser products on the market today offer a super-pulsed emission mode for this reason.

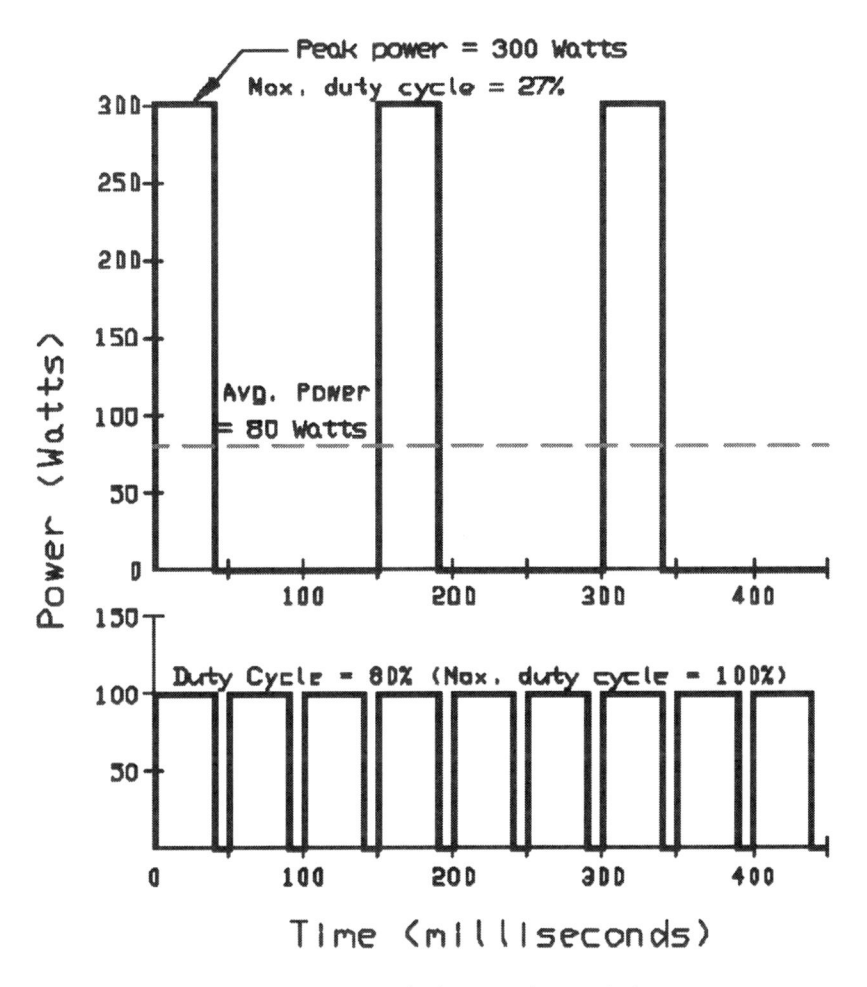

Figure 1-4 Superpulsed vs. gated cw emission

Pulsed modes: free-running, Q-switched, mode-locked

Solid-state lasers can typically be operated in one or more different types of pulsed emission modes. These differ from each other, and from quasi-CW lasers, by the range of pulse durations and peak power levels that are generated. Directly or indirectly, these differences in peak power and pulse duration afford different capabilities as a surgical or therapeutic laser modality.

Medical Lasers

So-called *free-running* or *normal-mode* pulsed lasers typically emit pulse durations of 0.1 to 10 milliseconds, and have maximum peak powers of about 1,000 to 20,000 watts depending on the product. Pulse duration may be user-selectable. Examples of free-running pulsed lasers used in laser medicine include holmium surgical lasers and pulsed Nd:YAG lasers used in laser dentistry. Pulse rates of 1 to 100 Hz are typical for free-running pulsed medical lasers.

There are two basic types of *Q-switched* lasers: electro-optically (EO) switched and acousto-optically (AO) switched. EO-switched lasers typically have pulse durations of 1 to 50 nanoseconds, whereas AO-switched pulse durations are usually 100 to 200 nanoseconds (for lamp-pumped lasers). In most cases, pulse duration is fixed for any specific Q-switched laser product and the user cannot select different pulse durations. Actual pulse duration may vary somewhat depending on the selected pulse energy and average power (pulse rate) setting. Pulse rates of 1 to 50 Hz are typical for EO-switched medical lasers, whereas pulse rates of 1,000 to 25,000 Hz are typical for AO-switched lasers.

Examples of EO-switched lasers include pulsed Nd:YAG lasers used to perform posterior capsulotomies and Q-switched ruby lasers for removing tattoos. The frequency-doubled Nd:YAG laser (often called the KTP laser) is an example of an AO-switched laser. Even though the frequency-doubled Nd:YAG lasers used for laser surgery are, strictly speaking, pulsed lasers, pulse rate is high enough (25,000 pulses per second), and peak power is low enough, that they are used much as a CW laser would be used in most medical applications. (This is why many laser surgeons think of the KTP laser as a CW laser).

Some solid-state medical laser products are operated in a *pulse-stretched* mode, which is a special type of Q-switched operation. Basically, the Q-switched laser pulse duration is increased from 10 or 100 nanoseconds to about 1 microsecond (1000 nanoseconds). This may be accomplished electronically or optically. In either case, the "stretched" pulsed duration allows one to deliver more pulse energy through an optical fiber delivery system than if shorter pulses were used.

Mode-locked lasers typically generate picosecond or femtosecond duration pulses. It is thought that such *ultrafast* or *ultrashort* pulses may provide

Medical Lasers

advantages over much longer nanosecond-pulse (Q-switched) lasers in some situations. For example, it may be that a much shorter pulse allows one to perform laser eye surgery with much less energy, possibly resulting in enhanced surgical precision and control. Mode-locked lasers are being investigated for corneal surgery and related ophthalmic procedures.

Figure 1-5 schematically compares pulse durations and peak power levels of typical free-running, Q-switched, and mode-locked laser pulses used in laser medicine. Super-pulsed and chopped mode pulses are also shown. Note that the horizontal and vertical axes are logarithmic (powers-of-ten) scales. For example, -6 on the horizontal axis indicates a pulse duration of 10^{-6} seconds = 1 microsecond, which is a shorter pulse duration than -3 = 10^{-3} seconds = 1 millisecond. Also on the horizontal axis, "0" = 10^0 = 1 second, and "1" = 10^1 = 10 seconds. On the vertical scale, a 6 indicates a peak power of 10^6 watts = 1,000,000 watts, which is higher than 3 = 10^3 watts = 1000 watts, for example.

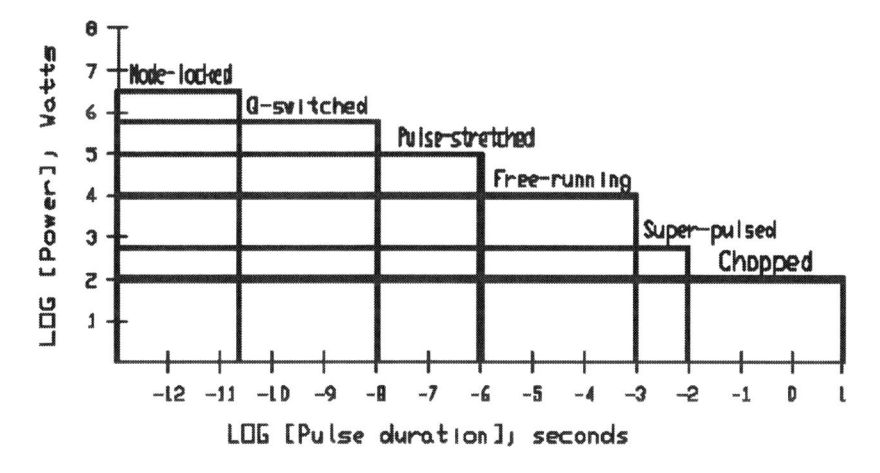

Figure 1-5 Comparison of medical laser pulse durations

In general, pulse energy decreases as shorter pulse durations and higher peak power levels are used. Assuming one uses enough energy to produce the desired tissue effect, smaller pulse energies typically allow for increased surgical precision or less damage to surrounding normal tissue. It may be counter-intuitive for some that a 1 million Watt (peak power) laser can be used to perform precise surgery. This is possible by virtue of the very short pulse durations, and (therefore) small pulse energies, that are employed.

Semiconductor diode lasers

Like transistors and related electronic components, diode lasers are solid-state devices made from semiconductor crystals. Similar to the light-emitting diode (LED) display components used in many electronic appliances, diode lasers emit light directly when an electric current passes through them. However, diode laser light is generated in a beam that is much more directional and monochromatic than the emission from an LED. This enables diode laser light to be focused to the very small spot diameters needed for medical applications.

Semiconductor diode lasers are "pumped" electrically by passing an electrical current through them. This feature of semiconductor diode lasers distinguishes them from all other types of solid-state lasers and is one reason why semiconductor diode lasers are referred to separately. The term "solid-state laser" is usually meant to indicate something other than a diode laser, even though diode lasers are in fact solid-state devices.

The active medium in a diode laser is basically a very small rectangular chip of semiconductor material with millimeter and sub-millimeter dimensions. Rather than having physically separate mirrors, most diode lasers incorporate the laser resonator micro-mirrors directly onto the ends of the semiconductor chip itself. Like other lasers, diode lasers require a specialized electronic power supply, controller, and cooling system. However, because diode lasers are electrically very efficient, their power supplies and cooling systems are typically much smaller, lighter, and more compact than for other laser systems of similar power capability.

There are many different kinds of semiconductor diode lasers, and they are often classified according to the composition of the semiconductor material used to make them. Diode laser semiconductor materials include aluminum gallium arsenide (AlGaAs), indium gallium arsenide (InGaAs), indium gallium aluminum phosphide (InGaAlP), and a variety of other combinations of gallium, indium, arsenic, aluminum, or phosphorus atoms. Different materials provide different laser performance capabilities, but are usually selected on the basis of the laser wavelength needed for the intended application. Considering the various research and industrial diode laser products available commercially, diode laser wavelengths span most of the range between 400 nm and about 25 microns (25,000 nm). However, any one device provides only a relatively small range of wavelengths. For the most part, the high power diode lasers used in laser medicine, and to pump solid-state lasers, are aluminum gallium arsenide (AlGaAs) lasers that provide a nominal laser wavelength of 800 nm, or indium gallium arsenide (InGaAs) lasers that provide wavelengths around 980 nm.

High-power semiconductor diode lasers (diode lasers) are already having an impact in laser medicine and surgery and will have an even bigger impact in the future. These lasers are very efficient devices electrically, converting as much as 50% of their electrical input power into laser output power. Just as importantly, high power diode lasers should become relatively inexpensive as production volumes increase.

Increased laser output efficiency often translates directly into important ergonomic and economic benefits. The diode lasers used in ophthalmic laser products are much more efficient than argon-laser alternatives resulting in a device that is more compact and lightweight by a factor of 3 to 5. Cooling and electrical service requirements are also greatly reduced, allowing operation from a standard 110 Volt outlet, or a battery, rather than a high-current 208 Volt electrical outlet. Laser portability is enhanced dramatically. Numerous other medical applications of diode lasers are being developed that might also benefit from highly compact, portable, and reliable lasers.

The diode lasers used in ophthalmic products can be purchased in quantity for about $1,000 each (some products require more than one however). Prices have been dropping as a result of the expanding use of diode lasers for medical applications, research applications, and as pump sources for solid-state lasers and fiber lasers.

The AlGaAs and InGaAs diode lasers currently used in dental products typically generate 2 to 10 watts depending on the product. All products can operate from a standard 110 VAC outlet. Electrical current requirements are low enough that a dedicated outlet is not needed, and some products can operate from a battery. Products are air-cooled, lightweight (10 to 20 pounds), and highly portable.

Diode-pumped solid-state lasers (DPSSL's)

Most of today's solid-state medical laser products employ arc lamps or flashlamps for optically pumping the active laser medium. Electrical input power to the lamp is converted into white-light pump power with about 50% efficiency, which is quite good. However, laser output efficiency suffers dramatically as a result of energy losses that occur as pump light energy is coupled into the dopant laser ions. The low overall efficiency of most lamp-pumped solid-state lasers, which ranges from less than 1% to about 5%, results primarily from such pumping-related losses. For example, a "typical" Nd:YAG laser that consumes 1000 watts of electrical power converts about 500 watts into pump light power. Of this, only 80 watts is actually coupled into the neodymium (Nd) laser ions. The other 420 watts of pump light is absorbed by the pump cavity and lamp and converted into waste heat. Of the 80 watts of pump power absorbed by the dopant ions in the laser rod, another 50 to 60 watts is dissipated as heat within the laser rod itself. Most of the remaining 20 to 30 watts emerges as output power if the laser resonator is designed properly.

When it is possible to use them, chromium-sensitized host crystals (described earlier) offer a partial solution to this pumping-efficiency problem. One can couple a higher fraction of pump light energy into the laser ions and improve laser output efficiency, but usually only with the tradeoff of increased thermal lensing in the laser rod. Depending on how much more thermal lensing there is, laser output gains can be offset completely if one considers how much power can be delivered through a fiberoptic delivery system.

Pumping a solid-state laser with an appropriate semiconductor diode laser is, in principle, a more efficient and straightforward way to pump a solid-state laser, and a more general and widely applicable solution to the problem of optical pumping efficiency. As implied by the foregoing discussion, diode-pumped solid-state lasers can be much more efficient than their lamp-pumped equivalents. By matching the diode laser emission wavelength to the *absorption wavelength* of the dopant laser ions, one can increase the proportion of pump light power that is coupled into the laser ions while reducing the amount of heat dissipated in the laser crystal. In other words, one can increase laser output efficiency and reduce thermal lensing relative to the situation of pumping with a lamp. However, this assumes that a high-power diode laser is available that emits a well-matched wavelength that can be absorbed by the dopant ions. Diode-pumping techniques are becoming more practical as new wavelengths and higher output powers become available from diode lasers, and as diode laser prices drop.

As an example, diode-pumping of a CW Nd:YAG laser can improve overall electrical efficiency by a factor of 3 to 5 (2 to 3% for a lamp-pumped laser, to 10% or more for the diode-pumped version). The physical size and power handling capability of the required laser power supply and cooling system are reduced by a similar factor. Reduction in size and weight of the laser resonator, power supply, and cooling system in turn leads to a laser system that is physically much smaller, more compact, and lighter. Replacement of lamps with diode pump lasers results in systems that are inherently more rugged and reliable, and that require less maintenance (no lamp replacement). Assuming that diode laser pump sources will someday cost about as much as a flashlamp or arc lamp, smaller power supplies and cooling subsystems should result in considerable savings in laser manufacturing costs and in lower purchase prices for the end-user.

Besides the end-user benefits of reducing the size and cost of medical lasers, diode-pumping should have an important impact by enabling new types of solid-state lasers. These will likely include lasers based on completely new solid-state laser materials, new combinations of existing solid-state lasers, and existing lasers with expanded performance capability. Diode-pumping techniques will allow one to use new solid-state materials that cannot be "lased" at all using arc lamps or flashlamps, thereby rendering new laser wavelengths accessible. New products that combine several solid-state lasers in one small package, at a fraction of today's total costs, will be more versatile pieces of capital equipment that pay for themselves in a shorter period of time. Along a similar vein, diode-pumping should also expand the performance specifications and application versatility of single-laser products by increasing the range of selectable output powers, wavelengths, and temporal emission modes. For example, diode-pumped holmium lasers can be operated at room-temperature in a continuous-wave (CW) or a pulsed temporal emission mode, whereas lamp-pumped versions can only be operated in a pulsed mode at room temperature.

Ideally, one would like to have the conveniences afforded by a diode laser for just about any medical application. However, it will be a long time, if ever, before diode lasers are available that can directly provide the various combinations of laser emission parameters needed for most of today's therapeutic laser applications, not to mention tomorrow's. Until such time, incorporating diode lasers into diode-pumped solid-state lasers appears to be a way to extend the performance capabilities of diode lasers without overly compromising the important ergonomic and economic benefits offered by diode lasers.

Important areas where diode-pumped solid-state lasers (DPSSL's) offer advantages for medical applications, relative to standalone diode lasers, include enhanced wavelength versatility, emission of very high power bursts or pulses of laser energy, and easier or more efficient coupling of laser energy into small-core optical fibers.

Fiber lasers

A fiber laser is a special type of diode-pumped solid-state laser. Instead of a bulk crystal, laser ions are doped into a "single-mode" silica or fluoride glass optical fiber having a very small core size (5 to 20 microns). The doped fiber is optically pumped with a suitable diode laser source to get laser emission at a wavelength determined by the dopant ion(s). The small fiber core size enables high-power fiber laser emission that exhibits very high (TEM_{00}) beam quality.

It is relatively easy to make high quality optical fibers doped with laser ions such as Nd, Yb, Er, Ho, and Tm (and combinations thereof) compared to trying to grow bulk crystals of high optical quality. Accordingly, fiber lasers can be realized at a wide variety of different wavelengths just by changing the combination of dopant ion(s) and glass fiber material. High-power fiber lasers that provide 10 to 100 watts of cw power at wavelengths such as 1000 nm, 1300 nm, 1500 nm, 1900 nm, 2100 nm, and 2700 nm may soon be available commercially.

Unlike other solid-state lasers, high power fiber lasers typically do not need an internal water-cooling loop to remove waste heat from the active laser medium. This is because absorbed diode-pump power can be distributed along the entire length of the fiber laser (which may be several tens of meters long coiled up in a 6"-diameter coil, for example), thereby facilitating waste heat removal using conductive or other passive cooling methods. Eliminating the need for an internal water-cooling loop results in a laser system that is lighter, more compact, more affordable, and more reliable.

Basic aspects of medical laser products

The ultimate purpose of surgical and other therapeutic lasers is to generate a laser beam that can be used to control therapeutic effects in tissue; that is, to **treat tissue**. The practitioner selects wavelength, temporal emission mode, (peak and average) output power, pulse energy, and perhaps divergence of the **treatment beam** according to the needs of specific treatment situations. In general, any one type of laser provides only a limited number of different combinations of treatment parameters so the surgeon may need to select among different types of lasers to tailor treatment parameters.

In addition to the treatment or "main" beam, most medical laser products also emit a low-power visible **aiming beam** to indicate where the treatment beam will impact on tissue when the laser is activated. This aiming beam is usually generated by a separate laser source but may also be generated by a lamp of some sort. The aiming light source is located in the same cabinet or electronic enclosure as the treatment laser.

Once the treatment beam parameters have been selected, the application of laser energy to tissue is usually controlled with a foot-controlled switch, or **footswitch**, so both hands are free during surgery.

A laser's **delivery system** refers to the opto-mechanical hardware used to direct the laser beam to the treatment site on or inside the patient. One end of the delivery system is attached to the laser. The other end is allowed to move freely so it can be manipulated by the practitioner. Basic types include articulated-arm and optical fiber delivery systems, which are discussed further in Sections 3 and 4. When functioning properly, the delivery system delivers the treatment and aiming laser beams so that they are spatially overlapped on the tissue surface at all times. **Coaxial cooling** may be required to prevent damage to the delivery system, and is usually accomplished by flowing gas (air, carbon dioxide, or nitrogen) or fluid (water or saline) through or coaxial with the delivery system.

As mentioned earlier, lasers generally consist of laser resonator, power supply, cooler, and controller subsystems. Most medical laser products have these packaged into a single roll-around cabinet or enclosure along with the aiming beam light source. Some products may incorporate the laser resonator and other optics into a separate "laser head module" that can be positioned closer to the patient. The laser head is then connected to the rest of the laser system via an electronic "umbilical" cable.

The **controller** subsystem automatically performs a number of tasks that would otherwise have to be performed by the operator. Besides controlling the laser power supply, resonator, and cooler to provide the laser emission parameters selected by the operator, the controller typically 1) **monitors** system status and performance and shuts the laser down in the event of malfunction, 2) properly **initializes** various laser settings and turns subsystems on in the proper sequence at laser turn on, and 3) **records** laser exposure data during a treatment session for later reference (many laser products have a thermal printer for outputting a hard-copy record). Many of today's medical laser products are microprocessor-controlled.

Almost all medical laser products today are air-cooled devices. Diode-based products are usually *"completely" air-cooled* and do not have an internal water-cooling loop inside the laser system. On the other hand, lamp-pumped lasers that are *"air-cooled"* usually do have an internal water loop. Like a refrigerator, these air-cooled lasers dump their waste heat into the room through a liquid-to-air heat exchanger. No tap water cooling services are required. Types of air-cooled lasers include **forced air-cooled** and **convection-cooled** lasers. Forced-air products incorporate an internal cooling fan that circulates air through the system. Convection cooled products do not use a fan and often run quieter as a result.

Water-cooled lasers incorporate a liquid-to-liquid heat exchanger and require external tap water cooling connections to carry heat away from the laser. Water-cooled lasers do not void their waste heat into the room environment, and may be preferred in situations where high laser powers are used for long periods in poorly air-conditioned rooms. Water-cooled medical lasers are very rare in this day and age.

Most medical lasers operate from and plug into an electrical wall outlet. Electrical service requirements vary considerably among products. High power lasers (up to 100 watts output power) usually require **single-phase** 208/220 VAC electrical service, although some products require **three-phase** 208/220 VAC service. (In essence, the terms "single-phase" and "three-phase" refer to the number and type of prongs on the power-cord plug, which ultimately determine where the laser can be plugged in). Amperage (electrical current) requirements can be as high as 50 Amps for a high power laser and as low 1 to 2 Amps for a low power device. Some medical lasers can be operated from a standard 120 VAC, 20 Amp electrical outlet. Some low power lasers used for ophthalmic surgery can operate from a battery.

Dedicated electrical service means there is only one wall outlet per electrical circuit so only one device can be operated at a time from that circuit. Except for lasers that have low current requirements (few Amps) some hospitals may require that a laser be operated only from a dedicated wall outlet, so that, should the circuit breaker that protects the circuit be tripped, other equipment in the operating room will not be affected.

Portability facilitates multi-specialty usage and increases the utilization of medical laser products. Lasers that can move between, and be used in, multiple operatories are more likely to be highly utilized than those that can't. Factors that affect portability include cooling requirements, electrical requirements, and product size and weight.

Air-cooled lasers are generally considered more portable than water-cooled lasers, as the latter can only be used in operating rooms with specialized tap-water cooling services and connections. Virtually all of the new medical lasers in recent years have been air-cooled products for this reason. A typical set of service requirements for a water-cooled laser might be an inlet water temperature of 35C, water flow rate of 2 to 4 gallons per minute, and a pressure of 40 psi or higher. A drain must also be provided to carry cooling water out of the operating suite.

Specialized electrical service requirements can also limit portability. Lasers that need three-phase service are generally considered less portable than single-phase lasers. Most of the newer products operate from a single-phase electrical outlet as a result. Current (amperage) requirements are also an important issue. Outlets that provide 30 amps of current at 208 Volts AC (VAC), or 220 VAC, are probably more common in most institutions than 208/220 outlets that provide more than 30 amps. However, the availability of various types of electrical outlets varies considerably between institutions. The prospective medical laser buyer should carefully evaluate where outlets appropriate for the product being considered do and don't exist. Installation of specialized electrical services is a frequently overlooked cost of acquiring new laser equipment. These costs can sometimes be avoided by selecting an equivalent product that operates from existing outlets.

Lasers that operate from a "standard" 120 VAC electrical outlet may not be as portable as one might think. Such lasers can be less portable than 208/220 VAC lasers if the buying institution's policies require operation from a *dedicated* 120 VAC outlet. Dedicated 120 VAC outlets may be less common in some hospitals and surgicenters than dedicated 208/220 VAC outlets.

Another portability issue relates to coaxial cooling of delivery systems. Some lasers have on-board gas or fluid canisters that install into the laser, and move around easily with the laser. Others are used with external cylinders or services that must be connected to the laser and that may impede portability.

Although less of an issue than electrical and cooling service requirements, large product size or weight can have a negative effect on portability (in spite of the fact that products roll around on casters). Multi-specialty usage may be discouraged if a product is too big and heavy to be moved *easily* between suites, or to a position *conveniently* near the operating table.

Diode lasers represent the ultimate in portability. Products are typically lightweight (30 pounds or less), compact, and plug into any 110V outlet. Electrical current requirements are low enough that a dedicated 110V outlet is not needed even for the highest-power medical diode laser systems.

Laser-Tissue Interactions, Effects, and Therapeutic Applications

A basic discussion of how lasers interact with biological tissue is provided for the reader who is new to the field of laser medicine and surgery. General aspects of how laser-induced tissue effects are used to provide therapy are also discussed. The section concludes with a description of various types of biological tissue and their features thought to be important for controlling laser-tissue interactions. We do not intend this to be an exhaustive treatment of the subject. Instead, we focus on issues and concepts needed to better appreciate new clinical and laser product developments.

Laser therapy and procedures

Laser-tissue interactions are physical processes used to create and control laser-induced tissue effects, which in turn are designed to provide therapy for the patient. Before getting into technical details, we first describe, in general terms, how lasers are used to provide therapy and why lasers are used when they are used. Possible future therapeutic uses are also described. Note that not all of the laser therapies described below are FDA-approved at this time; some are experimental.

Surgery Abnormal, pathological, or otherwise undesirable tissue is removed with a laser or with the help of a laser. When a laser is used, it is usually because it provides a level of ***surgical precision*** or ***surgical control*** not available with mechanical or other surgical modalities, or because the laser provides reduced bleeding during surgery. Many laser surgery procedures are really ***laser-assisted*** procedures that use a laser in conjunction with other modalities so as to take advantage of the laser's strengths and avoid its weaknesses. Lasers are used to perform ***open*** surgical procedures on external tissue structures and internal structures through long incisions, and to perform minimally-invasive surgery (MIS) procedures.

Endoscopic (minimally-invasive) surgery   Surgery is performed on internal organs and tissue without a major incision into the body, working through natural body orifices or through small skin incisions. The elimination of long incisions reduces patient trauma and speeds patient recovery. Important reasons why lasers are having an impact is that they a) allow one to design and implement ***endoscopic alternatives*** to highly traumatic open procedures, or b) they ***enable*** endoscopic ***surgical therapy*** where open surgery would be impossible or too risky to be justified.

Of the lasers that can be used endoscopically, lasers whose energy can be delivered through very thin and flexible optical fibers are having the biggest impact on endoscopic laser techniques. Fibers with overall diameters less than 1 millimeter permit very small incisions to be used or none at all (through natural body openings). In general, an important reason why a laser is used for a particular endoscopic procedure, rather than some other surgical modality, is because the laser can be used with a very thin optical fiber to perform less-traumatic surgery.

To be precise, the term "endoscopic procedure" implies that one uses a physically thin optical telescope device, called an ***endoscope***, designed to be inserted into the body through a body opening or small incision (see Section 3). The endoscope's function is to allow the surgeon to visually inspect tissue before and after surgery, and to control surgical instruments during surgery. We adopt a looser definition of "endoscopic" to include minimally invasive procedures performed with any one of several possible visualization or guidance modalities (see Section 3). We use the terms "endoscopic" and "minimally invasive" as synonyms.

Lithotripsy   The use of lasers to fragment kidney stones, gallstones, and other calculi, so that stone pieces can be excreted or removed endoscopically with mechanical instruments, is a special case of endoscopic laser surgery. In most instances where a laser is used, a ***pulsed laser*** is required that is quite different from lasers used for surgical procedures on soft tissue. Laser lithotripsy is typically used only where other non-invasive or less expensive minimally-invasive lithotripsy methods have failed, or are inadvisable.

Angioplasty The use of lasers to remove atherosclerotic plaque and recanalize blood vessels is another special case of endoscopic or minimally invasive laser surgery. As in the case of lithotripsy, laser angioplasty systems and fiber catheters tend to be quite different from those used for conventional soft-tissue laser surgery. Laser fiber catheters must be small in diameter so that they can be inserted into the blood vessel. They must be very flexible so they can negotiate the tortuous vessel passageways between the insertion point and the actual plaque lesion or blockage. At the same time, catheters must be capable of making a relatively large hole (2 to 3 mm in diameter) through the plaque lesion.

Lasers are sometimes used for angioplasty procedures because they provide technical or clinical advantages relative to the other angioplasty modalities (balloon angioplasty, other mechanical devices, and microwave catheters). Potential advantages include an ability to open up complete as well as partial blockages, and to more effectively remove calcified plaque lesions. Actual laser advantages have yet to be fully documented and widely accepted; situations where lasers are best used are still being investigated.

Cancer therapy For some types of cancer, ***photodynamic therapy*** methods that employ low power lasers and injectable organic dyes can be used to selectively kill malignant tissue. **Laser hyperthermia** procedures take advantage of the fact that tumors are more heat sensitive than normal tissue and use lasers to gently heat and selectively kill tumors. Both of these cancer therapy methods are relatively atraumatic for the patient and are being evaluated as possible alternatives, or adjuncts, to chemotherapy, radiation therapy, or surgery.

Tissue welding / fusion Lasers are used to seal or fuse biological tissue without sutures or with a significantly reduced number of sutures. Laser tissue welding may provide improved clinical outcomes in procedures where sutures are currently used, and may permit effective joining of very small tissue structures difficult or impossible to fuse with sutures or other mechanical closure devices.

Pain control   Lasers are sometimes used in dentistry to provide laser-induced analgesia (temporary desensitization of teeth without needle injections of local anesthetic). Laser methods are also being investigated for treatment of chronic pain in situations where drugs or electrical nerve stimulation are habit-forming or lose their effect after a period of time. Conventional surgical treatment, which involves cutting the indicated nerves with a scalpel or other mechanical instrument, can be imprecise and may lead to inadvertent impairment of sensory or motor functions. Some lasers may be able to perform such surgery with reduced neurological deficits.

Biostimulation   It may be that properly controlled low-power laser radiation can stimulate or otherwise alter metabolism of living tissue in therapeutically useful ways. Stimulation of wound healing is a possible example. Biostimulation is also referred to as **low-level light therapy (LLLT)**.

Basic types of laser-tissue interactions

According to present scientific understanding, laser radiation, or light energy, must be converted into some other form of energy to produce therapeutic tissue effects. The atoms and molecules that comprise biological tissue are ultimately responsible for *absorbing* laser radiation and converting it into other energy forms. Laser-tissue interactions are often categorized according to whether laser energy is converted into heat, chemical energy, or acoustic (mechanical) energy.

Photothermal   Laser light is absorbed by tissue and converted into heat energy resulting in a rise in tissue temperature. Where infrared wavelength lasers are used, the water component of tissue usually plays a dominant or important role in the absorption of laser energy. Water is heated directly with laser energy, and other molecules and tissues are indirectly heated via heat conduction. Other tissue molecules and components may also absorb at specific infrared wavelengths and play an important role in the tissue heating process. Visible wavelength lasers are absorbed poorly by water and usually rely on blood (hemoglobin) or other endogenous tissue pigments to absorb laser light and convert it into heat. Naturally occurring molecules that absorb visible wavelengths include hemoglobin, xanthophyll, and melanin. Protein molecules, DNA, and RNA absorb ultraviolet wavelengths strongly and usually play a dominant role in converting UV light energy into heat.

Photodisruptive (photoacoustic)   Pulsed laser energy is converted into acoustic (mechanical) energy in the form of a shock wave, or high-pressure wave, which physically disrupts or breaks apart targeted tissue. Although laser wavelength plays some role, laser peak power, pulse duration, pulse energy, and beam focusing are usually the critical parameters for controlling such interactions.

Photochemical   Laser light is absorbed and converted into chemical energy. Chemical bonds in the molecules that comprise tissue are broken directly by laser light, and/or molecules are excited into a biochemically reactive state. Laser wavelength is usually a critical factor. Short ultraviolet wavelengths (193 nm for example) are typically needed to maximize chemical bond-breaking processes and minimize photothermal processes. Specific visible and UV wavelengths may be able to induce photobiochemical reactions.

Photodynamic   In general, photodynamic interactions employ light-absorbing molecules to produce a biochemically reactive form of oxygen in tissue, called ***singlet oxygen***, when activated by light of an appropriate wavelength. As such, photodynamic interactions might be considered to be a special kind of photochemical interaction. Specific visible and near-infrared wavelengths can be used to control photodynamic interactions depending on the light-absorbing molecule (photosensitizer) used to mediate the interaction.

Biostimulation   The use of low power lasers to provide pain relief, stimulate wound healing, or alter other biological processes is a controversial subject. This is due partly to the fact that the physical, biochemical, and physiological mechanisms involved are not yet fully understood, and partly because patient benefits can be subjective and hard to quantify. If their effects are real, biostimulation interactions could be some combination of photothermal, photoacoustic, and/or photochemical interactions, but even this has yet to be established for specific effects. For this reason, we discuss biostimulation interactions as a separate category.

Each of the above categories of laser-tissue interactions, and the useful effects they produce, are described in more detail below. However, we first describe some of the basic aspects of what happens when a laser beam impinges upon, or *irradiates*, a tissue surface.

Laser irradiation of tissue

Initially, we assume that a collimated laser beam (not diverging or focusing) with circular cross-section is directed onto a tissue surface and that the beam is perpendicular to the tissue surface. We also assume the beam has some known power density (watts per unit area) that is uniform throughout the cross-sectional area of the beam. It will be important to relax some of these assumptions later on to discuss some of the finer and more subtle aspects of laser-tissue interactions (laser beams and their terminology are discussed in Section 1).

Laser beam *power density* plays an important role in creating and controlling laser-induced tissue effects. A typical situation might have a laser beam that delivers 10 watts of power to tissue in a 2-mm-diameter beam. In such a case, the *irradiated spot* on the tissue surface is circular in shape and has a diameter of 2 millimeters. The area of the irradiated spot is 3.14 square millimeters or 0.0314 square centimeters. Power density at the tissue surface is 10 watts ÷ (3.14 mm^2) = 3.2 watts per square millimeter, or 320 watts per square centimeter (laser power density is usually expressed in terms of watts per square centimeter). As shown by this simple calculation, power density can be controlled by adjusting laser power or the diameter and area of the irradiated spot on the tissue surface.

As the laser beam irradiates the tissue surface, some fraction of the beam's power is *reflected* off and away from the tissue surface. Most of the incident power, however, crosses the tissue surface and travels into tissue in the forward or propagation direction of the laser beam. In virtually all of the applications described in this report, we are interested in what crosses the surface and penetrates into tissue. Reflected power is usually an issue only to the extent that it reduces the useful power density of the forward-directed laser beam. (Some beam delivery devices can actually recover, or recycle, some of the reflected power and redirect it into tissue). It should be noted, however, that if the reflected power density is high enough, reflected power can cause inadvertent laser effects at sites other than the deliberately irradiated area and serious complications for the patient. Possible adverse effects of reflected laser energy must always be considered when performing laser procedures, but we rarely discuss the reflected beam from here on.

As the laser beam penetrates into the tissue it defines a so-called ***irradiated volume***. This volume of directly irradiated tissue is important for various reasons that will be discussed in the context of specific laser-tissue interactions. The irradiated volume's size and shape depend on two basic physical processes that act to attenuate (reduce the power density of) the laser beam as it penetrates into tissue: ***absorption*** and ***scattering***.

Absorption and absorption depth

Absorption is the physical process by which atoms and molecules that comprise tissue convert light energy (laser energy) into other forms of energy such as heat, chemical energy, acoustic energy, and non-laser light (see below). As a laser beam penetrates into tissue, absorption removes a certain amount of energy per unit time, or power, from the laser beam, and converts it into other forms of energy at the same rate. In this way, the power in the forward-directed laser beam is ***attenuated*** or diminished as it moves forward into tissue. Assuming the beam's cross-sectional area (diameter) remains constant, the beam's power density is also attenuated with tissue depth. At some depth, which we will loosely call the ***absorption depth***, the beam's power density is reduced to a low value that cannot produce a significant biological effect.

The strength of the absorption process, and the absorption depth we have defined, depend primarily on laser wavelength and tissue type (*i.e.*, the specific atoms and molecules that comprise the tissue in question). For example, the absorption of visible laser wavelengths (400 to 700 nm) by water molecules is very weak and the penetration depth is very long (meters). For the water layer thicknesses encountered in medical laser applications (up to a few centimeters) the absorption of visible laser light by water is negligible and water is said to **transmit** visible light with high efficiency. In contrast, infrared wavelengths are absorbed very strongly by water and have absorption depths that vary between about 1 centimeter (10^{-2} meters) and 1 micron (10^{-4} centimeters) depending on the specific infrared wavelength being used. Blood absorbs blue and green visible wavelengths very strongly; absorption depths associated with these wavelengths in blood are very short.

Scattering

We use the term **scattering** to collectively refer to all physical processes that attenuate a laser beam by deflecting beam power into directions **other than the intended forward direction**. These include scattering of light by individual atoms and molecules, by aggregates of atoms and molecules (intracellular structures, cells, particles), and by any of the optical inhomogeneities that characterize biological tissue.

Scattering reduces laser beam power density by increasing its effective cross-sectional area as it penetrates into tissue. Assuming the beam is not "clipped" by any large opaque obstructions in tissue, the beam usually retains a circular cross-section as scattering increases its diameter. Although the total power in the beam is not diminished by scattering, as it is when absorption occurs, increasing beam diameter and area result in a diminishing **power density** as the laser beam attempts to move forward into tissue. Power density in the beam is eventually reduced to a low enough level, at a tissue depth we will refer to as the **scattering depth**, that no significant biological effect is produced. Strength of scattering and scattering depth also depend on laser wavelength, and the biological material involved, but in ways that are much different than how absorption depends on these factors.

Penetration depth

Scattering and absorption occur to some degree in all types of biological tissue. Together, they attenuate the laser beam and determine the depth to which the beam can induce a significant biological effect. We use the term ***penetration depth*** to denote that both processes have been considered. In some instances, one process is much stronger than the other and penetration depth can be approximated by the shorter of absorption depth or scattering depth.

The reader should note that we have linked the definitions of penetration depth, absorption depth, and scattering depth to a minimum power density ***capable of producing a biological effect***. Accordingly, the actual value for each of these depths may depend on the type of biological effect we are discussing. It is probably more appropriate to speak of a ***thermal penetration depth*** when discussing photothermal interactions, a photodynamic penetration depth, etc. Penetration depth may also depend on the amount of time the beam is applied to tissue, since the minimum power density capable of inducing a biological effect often depends on the length of time for which that power density is applied to tissue.

Irradiated volume

Laser beam spot size (diameter) at the tissue surface, and the penetration depth of the laser wavelength being used, together determine a quasi-cylindrical volume of tissue that is directly irradiated by laser energy. However, the actual ***shape of the irradiated volume***, where laser power density is higher than some minimum value needed to induce a biological effect, depends on the relative and absolute strengths of absorption and scattering processes, and, in some cases, on beam focusing parameters at the tissue surface. This ***irradiated volume***, or ***illuminated volume*** of tissue as it is sometimes called, is established instantaneously (at the speed of light) upon the arrival of the laser beam at the tissue surface.

The total mass of tissue contained within the irradiated volume figures heavily in the amount of laser power and energy needed to induce the desired tissue effect. The shape of the irradiated volume, primarily its maximum diameter and depth, determine the maximum ***surgical precision*** that can be achieved easily with a given laser modality in a particular type of tissue (see below).

There are two general cases we will consider. More than two are possible, but two idealized cases will serve the tutorial needs of this publication.

Case I: Strong absorption   Absorption depth is much shorter than scattering depth, and absorption depth is less than, or roughly equal to, the chosen beam diameter at the tissue surface. The irradiated volume is then cylindrical in shape, or disk-shaped if the depth of the irradiated volume is much smaller than its diameter. The diameter of the cylinder/disk is approximately that of the chosen beam spot size, and the depth of the cylinder/disk is comparable to the penetration depth associated with the laser wavelength being used and tissue being irradiated. To achieve a high level of surgical precision when cutting or vaporizing tissue with a laser, one generally strives for a Case I type irradiated volume by selecting a laser wavelength that is absorbed strongly.

Case II: Moderate scattering, weaker absorption   In this case, scattering depth is considerably shorter than absorption depth, but scattering depth is longer than a few millimeters. For our purposes, the overall penetration depth determined by scattering and absorption could be 1 to 2 centimeters or more. Scattering processes cause the diameter of the penetrating beam to increase quickly with distance from the tissue surface. The actual size and shape of the irradiated volume is more complicated than a simple cylinder or disk; some authors approximate it as a half-sphere with its equator in the plane of the tissue surface and a radius equal to penetration depth.

The irradiated volume of Case II may be desirable when coagulating larger masses of tissue or when performing selective photothermolysis, photodynamic, or laser hyperthermia procedures (see below). It is usually not desirable when trying to precisely cut, vaporize, or coagulate tissue because the lateral and axial extent (depth) of the irradiated volume is not limited to the immediate area of the irradiated spot as it is in Case I.

So far we have assumed a collimated laser beam at the tissue surface. Another important feature of Case II is that one can vary the depth and overall size and shape of the irradiated volume by adjusting the degree of beam divergence or focusing at the tissue surface. Sapphire contact tips and sculpted quartz fibers take advantage of this fact when they are used to reduce the depth of the irradiated volume, and thereby enhance the surgical precision of 1064 nm Nd:YAG lasers and 800 nm diode lasers.

Fluorescence

If laser wavelength is appropriate, some types of tissue can **re-emit** a portion of laser beam energy they absorb. Light is re-emitted in all directions, like a light bulb, and at wavelengths other than the laser wavelength. Re-emitted wavelengths are **characteristic of the tissue being irradiated**. This re-emitted light, or **laser-induced fluorescence (LIF)**, can be used to identify and distinguish between different types of tissue and to determine the physiological status of tissue. While LIF has no direct therapeutic application, its use as a real-time diagnostic or guidance tool is being investigated in numerous applications as a way to enhance the safety and/or effectiveness of laser and other treatments. The first dental products have appeared that use LIF to distinguish between healthy and carious enamel. The terms LIF and *fluorescence* are often used synonymously.

Photothermal interactions, effects, and therapy

Photothermal laser-tissue interactions heat target tissue with laser energy in some controlled fashion. Thermal effects that can be induced with a laser include local hyperthermia, coagulation, vaporization, and selective photothermolysis.

Tissue in the irradiated volume absorbs laser energy and converts it into heat. This causes a temperature rise in the directly irradiated volume. If laser energy is applied for a long enough period of time, heat conduction will cause a temperature rise in tissue outside the irradiated volume as well. In this way, thermal effects can be created indirectly well beyond the volume of tissue that is directly irradiated.

The discussions of photothermal interactions presented below assume that laser energy interacts directly with tissue via absorption in the irradiated volume, or indirectly with tissue via the conduction of heat outside the irradiated volume. Some of the observations and comments made in this subsection do not apply to the use of laser-heated hot-tip devices, such as sculpted fibers and sapphire contact tips, which heat tissue primarily via heat conduction from a hot fiber tip, rather than by direct absorption of laser energy in tissue. The use of hot-tip devices in laser surgery is described in Section 4.

Laser hyperthermia

Hyperthermia is the condition of elevated tissue temperature (normal temperature being 98.6°F or 37°C). The term is used most often to refer to cancer therapy procedures, but we use the term to indicate any heating of tissue that avoids tissue coagulation or vaporization (see below). Lasers are used to create and control localized hyperthermia in a number of investigational procedures.

Laser hyperthermia is being used experimentally to treat some forms of cancer. Tumor temperature is raised to and maintained at a temperature in the range of 42° to 45°C for some period of time, *e.g.*, 15 to 30 minutes. It is thought that activity of an enzyme called **collagenase** is stimulated on a local basis resulting in the destruction of tumor blood vessels and tumor death. (Collagen, which is broken down by collagenase, is an important structural component of blood vessels). Fairly large and deep irradiation volumes (Case II volumes) help to establish and maintain a uniform temperature profile throughout the tumor volume. Accordingly, deeply penetrating wavelengths in the 800 to 1064 nm range are often used to implement laser hyperthermia procedures.

Laser thermal keratoplasty

A very highly controlled form of laser hyperthermia has been used to reshape the human cornea as might be done someday to correct near-sightedness and other refractive disorders. In a procedure called **laser**

thermal keratoplasty, the cornea is heated until its curvature is changed slightly thereby changing its refractive power. A shallow penetration depth, on the order of the cornea's thickness, seems to be required to heat the cornea properly.

Tissue welding

The fusion of soft biological tissues with lasers (***laser tissue welding***) is being investigated as a way to replace conventional sutures in some procedures. The ends of tissue structures to be welded together are heated to a temperature of 70° to 80°C for a period of time until the ends become "sticky". Tissue must be heated uniformly so full-thickness heating is accomplished without significant tissue shrinkage or coagulation.

Coagulation, necrosis, and hemostasis

Coagulation refers to the irreversible damage that occurs when tissue is heated to temperatures of 50° to 100°C for a long enough period of time. Permanent denaturation of proteins results in the whitening or "blanching" of tissue much like the clear fluid in a chicken egg (albumin) turns white when boiled. Considerable tissue shrinkage also typically occurs.

Cauterization of blood vessels to control bleeding during surgery, or **hemostasis**, is a common application of laser coagulation. Blood vessels are coagulated and shrunk to the point where they are effectively sealed off. The size of the largest blood vessel that can be sealed with a given laser depends on the thermal penetration depth of the wavelength being used and other factors. For example, the application of some amount of mechanical pressure, so the vessel is collapsed while it is being heated, allows one to cauterize larger blood vessels. The large thermal penetration depths achievable with 1064 nm Nd:YAG lasers and 800 nm diode lasers can coagulate vessels as large as 3 mm in diameter.

One way to obtain hemostasis is to pre-coagulate tissue with a laser before cutting or removing it with a scalpel, forceps, or other mechanical instrument. Another way is to choose an appropriate laser wavelength for cutting (see below) so that "simultaneous" hemostasis is provided (cutting and cauterization with a single pass of the laser cutting probe). Laser wavelength and thermal penetration depth must be matched in some sense to the largest blood vessels to be cut and sealed to achieve good hemostasis.

The term coagulation implies that tissue is irreversibly damaged or killed. Once tissue has been coagulated, the body responds in some way during healing to remove dead tissue, usually by re-absorbing it or sloughing it. The laser practitioner typically has no direct control over these healing processes. This situation may or may not be a concern, but usually is when the coagulated tissue is in the wall of a vessel, duct, or organ. Sloughing of dead tissue during healing can result in perforation of the wall should the full thickness of the wall be coagulated inadvertently. Serious post-operative complications can result.

On the other hand, there are situations where tissue is deliberately killed via laser coagulation (coagulation necrosis) so the dead-tissue sloughing process will remove it from the body. Tissue removal via coagulation necrosis and sloughing may be desirable when removal by laser vaporization (see below) would be too time consuming, too difficult, or otherwise impractical. Sloughing may be preferable to mechanical removal in some cases, especially where mechanical removal would be too difficult, time consuming, or unsafe for the patient.

Vaporization and cutting

Laser **vaporization** is the process of removing tissue by converting it into a gaseous vapor, usually in the form of steam or smoke, and aspirating away the smoke with an appropriate suction device. One of the most important advantages of laser vaporization for minimally-invasive surgery is that, by first converting them into vapor, solid tissue masses can be removed easily through small incisions or small body orifices.

Tissue removal by laser vaporization lends itself well to endoscopic surgery. In many instances, a very thin laser delivery fiber and a thin suction cannula are all that's needed to remove tissue. Laser vaporization is particularly advantageous where relatively bulky forceps would otherwise be needed to grasp and remove tissue, or where it would be awkward to try to use a scalpel to cut tissue. Thin and flexible laser delivery fibers (see Sections 3 and 4) can, in some situations, provide better access to tissue and a relatively unobscured view of the surgery site compared to endoscopic surgery with mechanical instruments.

Because vaporization involves a change in the physical state of tissue, laser energy and power needed to vaporize tissue is much more dependent on the size of the irradiated volume than are coagulation and hyperthermia effects. Assuming that soft tissue is mostly water (and that no conduction of heat out of the irradiated volume occurs) approximately 2500 Joules are required to convert 1 cubic centimeter of water at 37°C into steam at 100°C. About 90% of this energy is needed just to convert liquid water at 100°C into steam at 100°C, and only 10% is needed to heat the water from body temperature to 100°C. Therefore, a substantial reduction in the amount of laser energy needed to vaporize soft tissue can be obtained by reducing the irradiated volume of tissue. This is usually accomplished by decreasing the laser spot size on tissue and/or by choosing a different laser wavelength that reduces the depth of the irradiated volume (that reduces thermal penetration depth).

Complete or true vaporization of hard substances such as bone and tooth enamel, which contain hydroxyapatite and other inorganic substances, requires heating of the inorganic component to temperatures much higher than 100°C if it is to be converted completely into a gaseous state. (Such inorganics typically have much higher melting and boiling/vaporization temperatures than water). However, complete conversion of all solid and liquid components into gaseous vapor rarely occurs in practice when "vaporizing" tissue with a laser, especially where hard tissue is involved. Instead, tissue is probably removed as a gaseous mixture of steam and very small solid particles. It is thought that the water component of tissue is rapidly heated to the point where it explosively turns to steam and mechanically disrupts tissue. Solid tissue is broken up (photodisrupted) into very fine particles that are entrained in, and carried away by, the expanding jet of steam. This mixture of steam and particles can usually be suctioned away easily. Photodisruption of tissue provides a possible explanation as to why some pulsed lasers are more effective than continuous-wave lasers for precisely ablating hard materials like bone and tooth enamel.

The "vaporization" process just described is thought to apply primarily to the use of infrared- and visible- wavelength lasers. The tissue vaporization process appears to be different for some UV wavelength lasers in that photochemical mechanisms play a larger role and photodisruption and photothermal mechanisms play a smaller role.

We rarely concern ourselves with such subtle differences when using the term "vaporization" in this publication. The basic point is that solid tissue is removed as gaseous vapor or smoke, usually with a suction device. We tend to use the term tissue "vaporization" as a synonym for tissue "ablation" (removal over an area), but we also use the term "vaporization" to indicate that a laser is being used rather some mechanical ablation instrument.

Laser **cutting** is a special case of tissue vaporization in which a thin linear or curved path is vaporized in tissue, *e.g.*, about 1 mm wide, so tissue edges can be separated mechanically. Laser excision of tissue (use of a laser to make cuts so target tissue can be removed mechanically as one solid chunk) is often a much faster process than vaporizing the entire tissue mass with a laser.

By virtue of the hemostasis that may occur simultaneously, cutting or vaporizing tissue with a laser can result in reduced bleeding compared to use of mechanical instruments. Tissue at cut or ablated tissue edges is coagulated over a distance that depends on a number of factors, including wavelength, power density at the outer periphery of the laser beam, number of passes made with the beam to make a full thickness cut, and the type of tissue being cut. In many situations, however, selection of an appropriate laser wavelength for the tissue involved is the first step toward achieving good hemostasis. Where a pulsed laser is being used, variation of laser pulse duration can sometimes be used to control hemostasis if the surgeon has the ability to adjust pulse duration over a suitable range.

Vaporization with pulsed vs. continuous-wave lasers

The main objective when vaporizing or cutting tissue with a laser is to heat the irradiated tissue volume until it vaporizes. However, **heat conduction** always counteracts one's efforts to raise the temperature of the irradiated volume. Heat conduction transfers energy out of the irradiated volume and into surrounding tissue thereby raising the temperature of surrounding tissue. Referring to our earlier example, much more than 2500 Joules of energy may be required to vaporize 1 cubic centimeter of soft tissue, or vaporization may be prevented altogether, if tissue is heated too slowly (laser power is too low) and heat conduction is given sufficient time to rob heat energy from the irradiated volume. The vaporization process is much less efficient in such a case and thermal injury to surrounding tissue is more extensive. Of course, some amount of heat conduction into surrounding tissue may be desirable. "Collateral" coagulation necrosis in surrounding tissue, due to heat conduction out of the irradiated volume, is one of the mechanisms responsible for hemostasis when cutting or vaporizing tissue with a laser.

Some very high precision, low hemostasis applications require cutting or vaporization with minimized thermal necrosis in surrounding tissue. To accomplish this, sufficient energy to vaporize the irradiated volume must be applied in a short enough period of time so that heat conduction transfers very little heat out of the irradiated volume. A pulsed laser may be required. The dimensions of the irradiated volume and the type of tissue involved determine a **thermal relaxation time** that, in essence, signifies how short the laser pulse must be in order to minimize heat conduction and collateral thermal necrosis. Each laser pulse must deliver the total amount of energy needed to vaporize the irradiated volume, and must have a pulse duration comparable to, or less than, the irradiated volume's thermal relaxation time. For the same pulse energy, significantly longer pulse durations (lower peak power levels) induce increasingly wider zones of thermal necrosis in surrounding tissue up to some limiting value. Lasers with variable pulse durations comparable to the irradiated volume's thermal relaxation time can be used to control collateral thermal necrosis and the amount of hemostasis obtained.

Surgical precision and control

Surgical precision refers to the surgeon's ability to treat tissue that he/she wants to treat (target tissue) and to leave in its original state surrounding normal tissue not targeted for treatment. Manually operated mechanical instruments, like a scalpel, produce no heat when being used and are generally accepted as the gold standard for surgical precision.

There is usually an important tradeoff to be made regarding surgical precision versus hemostasis. Hemostasis is the result of collateral thermal necrosis at the edges of cut or ablated tissue areas. Deeper or wider zones of collateral necrosis are required to achieve increasingly better hemostasis. On the other hand, sloughing of thermally necrosed tissue during healing can result in removal of tissue not intended for removal by the surgeon, thereby resulting in degraded surgical precision. In this way, deeper or wider zones of thermal necrosis necessarily result in reduced surgical precision. The surgeon's ability to do surgery near delicate anatomy may be compromised if delicate normal tissue lay within the collateral necrosis zone associated with the target tissue.

An important reason why lasers are useful surgical instruments is that they bridge the gap between monopolar electrosurgical instruments and mechanical instruments in terms of the surgical precision and hemostasis they afford. In general, lasers afford more precision and surgical control than monopolar electrosurgical and cryosurgical devices. To the extent that they induce some thermal injury, lasers are less precise than manually operated mechanical instruments (not including instruments like high-speed drills that can generate considerable heat in tissue and cause thermal necrosis).

The variety of laser wavelengths that are now available, especially in the mid-infrared wavelength region between 1000 nm and 3000 nm, allow one to leverage considerable control over thermal penetration depth/necrosis and surgical precision. Clinical laser products have already appeared that provide multi-wavelength capability as a way to adjust surgical precision. Lasers capable of operating in more than one temporal emission mode, *e.g.*, Q-switched or free-running pulsed, or that provide adjustable pulse durations, also allow one to tailor the level of surgical precision to the application at hand. The use of hot-tip fiber delivery accessories such as sculpted fibers and contact tips allow one to adjust surgical precision to some degree by interchanging fiber tips of various shapes.

Table 2-1 lists various surgical laser systems used to perform conventional soft-tissue surgical procedures and suggests where each might fall along the surgical precision/hemostasis continuum. (This listing is based on our judgement and may not be widely agreed upon). Actual ordering may also depend on the surgical procedure being considered. The listing assumes blood-perfused soft tissue which strongly absorbs green laser wavelengths (argon and frequency-doubled Nd:YAG). Although we have listed the precision of erbium and excimer lasers as being less than that of manually operated mechanical instruments, the difference can be insignificant.

Table 2-1: Relative surgical precision and hemostasis on soft tissue

PRECISION
Manual mechanical instruments
193 nm excimer lasers
Erbium lasers, 308 nm excimer lasers
Carbon dioxide lasers (continuous-wave)
Argon, doubled Nd:YAG, holmium, CW Nd:YAG with hot-tip fiber
CW Nd:YAG with bare fiber in non-contact mode
Monopolar electrosurgical
HEMOSTASIS

Another issue related to surgical precision is that of ***surgical control***. As we use the term in this report, surgical control refers to the ability to remove layer by thin layer of tissue. A typical example of very good surgical control is the ability to remove the shell of a chicken egg without causing any thermal or mechanical injury to the underlying soft tissue membrane. Perhaps an example of even better surgical control is to ablate only one-half or one-fourth of the way through the egg shell, and then stop, without fracturing the shell. Various types of pulsed lasers can in fact provide such high levels of surgical control that are difficult to reproduce with manually-operated mechanical instruments or continuous-wave lasers. In fact, pulsed lasers such as holmium, erbium, and excimer lasers can remove tissue in layers that are consistently only a few microns or tens of microns thick.

Pulsed holmium and erbium lasers are being investigated as alternatives to diamond-burr drills in some applications where a bony covering must be removed without injuring underlying nerve or blood vessel tissue. Excimer lasers and erbium lasers are being investigated in procedures that shave thin layers off the cornea of the eye in an effort to alter the refractive power of the cornea. For the most part, surgical precision and surgical control go hand-in-hand. Lasers that provide very good surgical precision also typically provide a level of surgical control that matches or exceeds that of mechanical instruments.

Lasers as bloodless scalpels

As the preceding discussions imply, lasers are used to do conventional surgery much like one uses standard mechanical instruments. Lasers can be used to ablate tissue, as with a curette or forceps, or to incise, excise, resect, dissect, or amputate tissue as one would use a scalpel. Some laser products allow one to use fiber delivery systems in direct contact with tissue so tactile feedback much like that of a scalpel is provided.

There are of course important differences. Perhaps the most obvious is that hemostasis can be provided depending on the type of laser being used and the type of tissue. Bleeding is only reduced in some cases, rather than prevented altogether, but usually enough that visualization of the surgery site is greatly improved. Operating time can be reduced as a result. The number of occasions where surgery has to be halted altogether due to uncontrolled bleeding may also be reduced.

Another practical issue that can arise relates to the use of pulsed lasers versus continuous-wave lasers for cutting. Used at appropriately high power, continuous-wave lasers cut soft tissue with a smooth and fast motion similar to that obtained with a scalpel. Pulsed lasers that provide sufficiently high average power at pulse rates of 100 pulses per second, or higher, can also cut tissue smoothly. However, some pulsed laser products have maximum pulse rates of only 10 to 20 pulses per second or less. Such low pulse rates make for a much slower and/or rougher cutting action, regardless of what the average power might be, and are less useful where extensive tissue cutting is required. Pulse rates of 30 to 100 pulses per second are much more useful and even higher pulse rates may be preferred in some situations.

In general, cutting or ablation speed is an important difference between lasers and mechanical instruments. As noted earlier, many laser surgery procedures are really *laser-assisted* procedures. Where mechanical instruments can be used, they can usually remove a given mass of tissue faster than even the highest power medical laser. Since operating time for any surgical procedure is an important consideration, lasers are frequently used only as an adjunct to mechanical instruments. The laser is used when it provides a useful advantage, such as better precision or control, better access to tissue, or better hemostasis, and is typically set aside when the only objective is to debulk tissue.

Selective photothermolysis

So far in our discussion of photothermal interactions we have described situations where laser pulse duration and wavelength were important for limiting tissue heating to that within the irradiated volume. Laser beam spot size was chosen to determine the area, and laser wavelength was chosen to control the depth and shape of the irradiated volume. When appropriate, a suitably short laser pulse duration was chosen to prevent or minimize the conduction of heat into tissue outside the irradiated volume.

Selective photothermolysis refers to the use of a properly selected laser wavelength and pulse duration to limit the heating of tissue to specific structures *within the irradiated volume itself*. By selecting a laser wavelength that is selectively absorbed by the target structure(s), and a short enough pulse duration that heat conduction from the target structure(s) into surrounding tissue is minimized, one can selectively heat the targeted structure(s) without causing thermal injury to nearby tissue within the irradiated volume. An important assumption here is that a laser wavelength can be used that is strongly absorbed by targeted tissue structures and weakly absorbed by surrounding tissue in the irradiated volume. A Case II irradiated volume is typically employed so that a large number of target structures and a large area or volume of tissue can be treated simultaneously. Selective photothermolysis was first described by Parrish and Anderson [see for example, *Science* 524 (1983)].

Where they can be used, selective photothermolysis (SP) methods offer useful advantages over other photothermal laser techniques. The need to precisely aim or focus the laser beam on the targeted tissue structure is eliminated. Microscopic as well as macroscopic structures can be treated easily, and multiple target structures in a fairly large area or volume of tissue can be treated simultaneously. Surgery can be performed on target structures lying well below the tissue surface without damaging overlying tissue.

In most cases, the objective of SP is to selectively heat and coagulate, vaporize, or otherwise destroy the targeted tissue structure(s), the remains of which are then absorbed during the healing process. Perhaps the best examples of where SP is used clinically are the treatment of port-wine lesions, tattoos, and pigmented lesions.

Pulsed dye lasers that generate a 585 or 600 nm wavelength, a pulsewidth of several hundred microseconds, and pulses energies on the order of 1 Joule are used to selectively coagulate the small swollen blood vessels responsible for port-wine stain lesions. These microvessels lay well below the skin surface and are heated and coagulated without thermally injuring overlying or surrounding tissue. Selective photothermolysis of these vessels typically reduces operating time, minimizes pain and discomfort during treatment, and reduces chances of scarring compared to use of other lasers with different wavelengths and pulse durations, or non-laser treatments.

Q-switched ruby lasers (wavelength of 694 nm and a pulsewidth of 50 nanoseconds) are used in similar fashion to remove certain types of tattoos without scarring. Subsurface tattoo pigment particles are selectively heated and destroyed. Pigmented lesions such as age spots, brown birthmarks, and freckles can also be erased with these lasers via selective photothermolysis of the melanin particle component of brown skin lesions.

Photochemical interactions and ablation

Photochemical interactions involve the conversion of laser energy into some form of chemical energy as the primary mechanism by which the intended therapeutic effect is controlled. Some amount of tissue heating may or may not also occur but is not utilized directly to produce the desired effect.

Photochemical ablation

Photochemical ablation is the process by which chemical bonds are broken directly with laser energy and tissue is ablated. As a result, tissue can be ablated and cut with virtually no thermal injury to surrounding tissue outside the irradiated volume, and often with exceedingly smooth surfaces at the cut edges. An appropriately short laser wavelength must be used to break chemical bonds in target tissue, and in most cases is an ultraviolet wavelength shorter than 250 nm. Such short wavelengths afford very shallow penetration depths in most tissues and can be focused to very small spot sizes. Tissue can be ablated very precisely and with a high level of surgical control to etch patterns with sub-micron spatial details. The clinical use of 193 nm excimer lasers to photochemically ablate and reshape the cornea, as a way to correct refractive disorders, is probably the best example of the extreme surgical precision and control that can be achieved via photochemical ablation.

The photodynamic interaction described below is a special type of photochemical interaction. Some of the biostimulation interactions and effects described below may also involve photochemical interactions.

Photodynamic effect and photodynamic therapy (PDT)

In most cases of interest, photodynamic interactions employ laser light and light-absorbing **photosensitizer** molecules to produce a biochemically reactive form of oxygen, called **singlet oxygen**, in tissue. Absorption of light excites the photosensitizer molecule into a more energetic molecular state. The excited photosensitizer molecule then transfers its energy to a nearby oxygen molecule (which cannot absorb the laser light directly) resulting in the formation of singlet oxygen. Singlet oxygen is cytotoxic and is thought to oxidize and permanently destroy some critical tissue component leading to localized tissue destruction.

As it is most frequently applied, **photodynamic therapy (PDT)** is a cancer therapy technique that uses a visible wavelength laser to locally activate singlet oxygen and kill tumors. An artificial organic dye/photosensitizer solution is injected into the patient's bloodstream and the photosensitizer selectively accumulates in tumor tissue over a period of hours or days. The tumor is then irradiated for some extended period of time (10 to 15 minutes or longer) and killed with low-power laser energy having a wavelength absorbed by the photosensitizer dye. Afterwards, the tumor degenerates and regresses over a period of several weeks.

Case II irradiated volumes are typically employed during PDT. Rapid and effective treatment of larger tumor masses requires a deeply penetrating laser wavelength so that all photosensitizer molecules retained by tumor tissue can be activated with a single application of light (single treatment session). Ideally, a photosensitizer is used whose activation wavelength can penetrate deeply so the entire tumor mass is encompassed within the irradiated volume of the treatment beam.

A photosensitizer dye called hematoporphyrin derivative (HPD), once marketed under the name **Photofrin**[tm] (Quadra Logic Technologies, Vancouver, BC), has been used in many clinical PDT studies. (Photofrin-based PDT procedures were the first to gain FDA approval.) A laser wavelength of 630 nm, typically provided by a laser-pumped dye laser, or red diode laser, is used to activate Photofrin.

Besides cancer therapy, a variety of other PDT treatments are being used clinically or being developed. Included in the list are treatment of age-related macular degeneration (AMD), severe psoriasis and venereal warts, and photodynamic inactivation of viral and bacterial contaminants in blood transfusion products.

Some PDT dyes *fluoresce* when illuminated with a low-power laser or other light source, which provides the basis for laser-based detection of tumors (see earlier discussion of laser-induced fluorescence). Tumors can be identified and localized that are not visible to the naked eye by illuminating a broad area with the laser. In most cases an image intensifier camera is needed to "see" the fluorescence. Some of the PDT laser systems being developed afford tumor detection and treatment capability in the same device.

Photoacoustic / photodisruption interactions, effects, and therapy

Pulsed lasers can be used to create acoustic shock waves (high amplitude pressure waves) in soft and hard biological tissue. These pressure waves can produce mechanical stresses sufficient to disrupt or break apart tissue. Laser energy is converted into acoustic energy, which is a form of mechanical energy. These are usually referred to as photodisruption, or photoacoustic, interactions and effects.

High peak laser power densities and relatively short pulse durations are usually needed to generate useful amounts of acoustic energy. Laser pulse durations from 60 picoseconds to several hundred microseconds have been used. Laser energy must either be focused to a small spot with a lens or delivered through a small-core optical fiber to efficiently convert laser energy into acoustic energy. An energetic acoustic pressure wave is generated at the laser beam's focal point, or at the fiber tip, and then travels away from this point at high speed (the speed of sound). This pressure wave expands in all directions (unless it is concentrated and directed with an appropriately designed acoustic reflector) so that the forces it generates diminish rapidly with distance from the generation point. Mechanical disruption effects can be limited to the immediate area of the generation point by limiting the amount of energy in the laser pulse.

The mechanism by which laser energy is converted into acoustic energy seems to depend primarily on laser pulse duration and peak power. The mechanism with very short pulsewidths, *e.g.*, 10 nanoseconds, often involves the creation of a so-called "laser-induced plasma" at the tissue surface. This plasma efficiently absorbs and converts laser energy into acoustic energy in tissue that may or may not strongly absorb the laser wavelength being used. As a result, photoacoustic tissue effects can be fairly consistent between different types of tissue. However, the amount of laser energy that can be delivered reliably through small-core fibers with such short pulsewidths is limited. Clinical photodisruption applications of these very short pulses are limited primarily to ophthalmic membranectomy procedures in which delivery of laser energy through an optical fiber is ***not*** required.

Lasers that employ a 1 microsecond pulsewidth, or longer, can deliver higher pulse energies through optical fibers. However, these longer pulse durations seem to require strong absorption in target tissue to achieve efficient energy conversion and useful amounts of acoustic energy. Photodisruption effects in different types of tissues are typically less consistent than for nanosecond or shorter pulses.

Membranectomy

Q-switched Nd:YAG lasers (1064 nm) are used to photodisrupt various types of pathological soft tissue membranes in the eye and in a completely non-invasive fashion. The 1064 nm wavelength is transmitted through the cornea, lens, and vitreous humor, with little or no absorption or attenuation, and with no thermal injury to tissue. The beam produces a photodisruptive effect only where the beam is brought to a focus with a slit lamp delivery system. Membranes are severed or punctured by focusing the beam on or close to the target structure, and firing one or more laser pulses. Photodisruption cuts the soft tissue with virtually no thermal necrosis in surrounding tissue. Ocular photodisruption applications typically use pulse energies of about 10 milliJoules, pulse durations of 10 nanoseconds or less, and focused beam spot sizes of about 10 microns. Research suggests that considerably less pulse energy is needed to produce the same effects and perform the same procedures if picosecond pulse durations are used.

Lithotripsy

Pulsed lasers are used clinically for fragmenting kidney stones, ureter stones, and gallstones inside the body so that pieces can be excreted easily or removed via minimally invasive surgery. Laser energy is delivered to the stone and fragmented with a small-core optical fiber. The pulsed dye lasers used for lithotripsy deliver about 100 milliJoules of energy in a 1 microsecond long pulse through a 300 micron core fiber. A wavelength is used (504 nm) that provides strong absorption in the stones most likely to be encountered, and weak absorption in the soft tissue of the ureter wall (so as to avoid accidental thermal injury to the wall). Solid-state Nd:YAG, holmium, and alexandrite lithotripsy lasers have also been used clinically.

Ablation of calcified tissue

Pulsed UV and infrared lasers can precisely ablate calcified tissue such as bone, tooth enamel, and calcified arterial plaque, probably because photoacoustic interaction mechanisms play an important role in the ablation process. Strong absorption of the laser wavelength being used by the water component or hydroxyapatite components of hard tissue also seems to be a prerequisite for precise ablation in most situations. In some cases, weakly absorbed laser wavelengths can achieve similar results if a laser-induced plasma can be produced at the tissue surface. Pulse durations of several hundred microseconds and pulse energies of a few hundred milliJoules are able to ablate hard tissue when used with a 300-um-core fiber.

Tumor disruption

Although such research is in a very early phase, there is some reason to believe that structural differences in the blood vessels (vasculature) of tumors may be more sensitive to the effects of laser-induced shock waves than normal tissue vasculature. Photodisruption could eventually prove to be another form of cancer therapy.

Biostimulation and pain relief

Research has been going on for some time now, mostly in Japan, the former Soviet Union, and Europe, which suggests that irradiation with a low-power laser can speed up the process of wound healing and similar regenerative processes. Other research indicates that low-power laser light can be used to relieve chronic pain as in arthritis. We collectively refer to such interactions and effects as **biostimulation**. The use of biostimulation effects is often referred to as **low-level light therapy (LLLT)**.

Historically, claims regarding the therapeutic effects of biostimulation have been highly controversial, at least in the USA. The biochemical and physiological mechanisms involved are not well understood. Implementing a well-controlled scientific study is very difficult because all of the experimental variables that affect the end result have yet to be identified. Another reason why it is difficult to prove anything scientifically is that end results are often subjective and hard to measure and quantify. Nevertheless, this situation is beginning to change. FDA has been granting an increasing number of clearances and approvals for LLLT applications and it seems likely more applications will emerge in the future.

Biomolecules, Tissues, and Materials

In this subsection, we describe certain molecules, materials, and tissues that often play an important role in laser-tissue interactions. Laser wavelengths that match the absorption and transmission properties of tissue are discussed to give the reader a better sense of why certain lasers, and not others, are chosen for surgery on particular types of tissue. We don't mention ultraviolet (UV) wavelengths explicitly in most cases, but it should be understood that UV wavelengths are absorbed strongly in virtually all types of tissue. As a result, most tissue can be cut or ablated precisely with most UV wavelengths. Ultraviolet wavelengths, especially those shorter than about 300 nm, are absorbed strongly by tissue proteins, amino acids, DNA, RNA, and other biomolecules.

Water

For our purposes, water does not absorb visible wavelengths (400 to 700 nm) to any significant degree and is said to transmit visible wavelengths with high efficiency. This fact is important when doing surgery in a wet or flooded surgical field where laser energy must be delivered to target tissue through an intervening layer of water. Infrared wavelengths shorter than about 1200 nm can be transmitted with reasonable efficiency through water layers less than 2 centimeters thick. In most cases however, one is usually much more interested in infrared wavelengths that water does absorb.

Water is present in significant amounts in all soft and hard tissue. Approximate weight percentages for the water component of various tissues are: skin (70%), aorta (79%), cartilage (75%), bone (10 to 30%), dentin (13%), and enamel (2 to 3%). Because water is so prevalent, absorption of laser energy by water can usually be relied upon to efficiently heat tissue if a strongly absorbed wavelength is chosen.

Laser-Tissue Interactions, Effects, and Therapeutic Applications

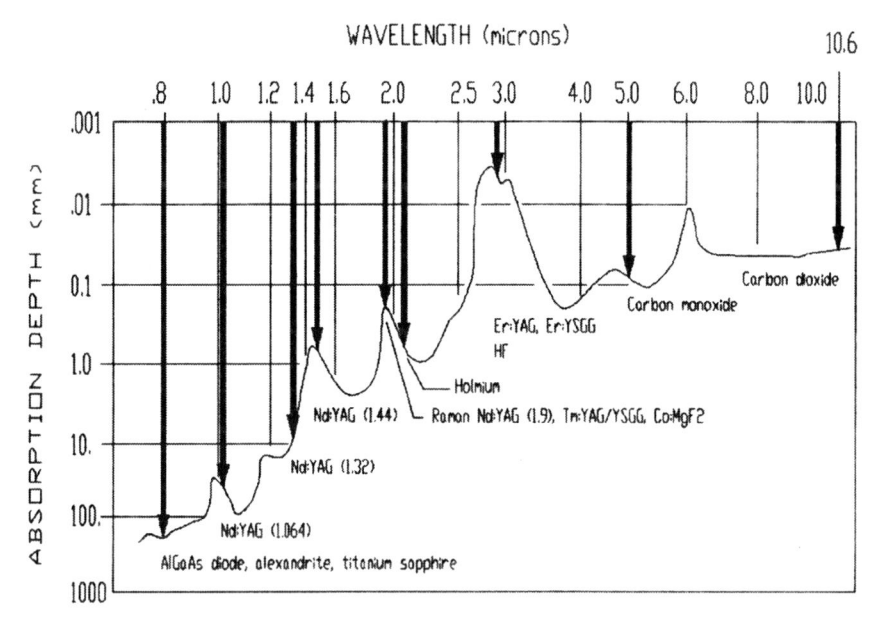

Figure 2-1 *Absorption depth in liquid water vs. IR wavelength (1 micron = 1000 nm = .001 mm)*

Figure 2-1 shows how the absorption of liquid water (as opposed to water vapor) varies with infrared wavelength. To be more precise, we show absorption depth as a function of wavelength. Absorption depth is defined here as the depth at which laser beam intensity is attenuated to 37% of its initial value at the water surface (depths at which the beam is attenuated to 1% are roughly 4 times those shown in the figure). Note that logarithmic rather than linear axes are used in the plot. Absorption depths corresponding to various infrared laser wavelengths are highlighted with arrows.

Of great interest for controlling laser tissue interactions are the several local maxima, or ***absorption peaks***, that occur in the mid-infrared wavelength region between 1200 nm and 3000 nm. These peaks occur at nominal wavelengths of 1450 nm, 1930 nm, and 2940 nm. Absorption depths corresponding to wavelengths at and around these peaks get progressively smaller as one proceeds to longer wavelengths and span a dramatic range of values from 10 millimeters (mm) down to .002 to .003 mm (2 to 3 microns).

The attenuation of infrared laser energy in actual tissue is a much more complicated process than what occurs in pure water. However, an expanding base of experimental and clinical data indicates that, for a given type of tissue, there is a good qualitative correlation between the minimum achievable penetration depth in tissue and absorption depth in water as infrared wavelength is varied. Wavelengths that penetrate less in water (have shorter absorption depths) can usually be expected to penetrate less in the target tissue than wavelengths that penetrate more deeply in water. Even this relatively crude qualitative correlation is proving useful for controlling laser-induced tissue effects.

Hemoglobin and blood

Hemoglobin is the molecule in blood responsible for binding, carrying, and delivering oxygen to tissue. It is almost as prevalent as water in tissue. Notable exceptions are cartilage and dental hard tissue such as cementum, dentin, and enamel. Hemoglobin gives blood its red color by reflecting red wavelengths and strongly absorbing blue and green wavelengths. Deoxygenated hemoglobin absorbs red wavelengths more strongly than oxygenated hemoglobin (oxyhemoglobin), which is why the deoxygenated blood in veins is much darker than the oxygenated blood in arteries.

Lasers that emit blue and green wavelengths are absorbed strongly by hemoglobin and can be used to do precise surgery on soft tissue that is well perfused with blood. These lasers include the argon laser (514 nm and 488 nm) and the frequency-doubled Nd:YAG laser (532 nm). However, the surgical precision obtained with these modalities can vary considerably between tissues as the degree of blood perfusion varies. Lasers that emit yellow wavelengths around 577 nm are also absorbed by hemoglobin, but not as strongly as blue-green wavelengths.

Blood consists of large amounts of water as well as hemoglobin. The water component absorbs a wavelength of 1064 nm about 10 times less strongly than 1320 nm. However, because oxygenated hemoglobin absorbs 1064 nm more strongly than 1320 nm, oxygenated whole blood (including the water) absorbs 1064 nm about a factor 3 times **more strongly** than 1320 nm. Therefore, the absorption depth of 1320 nm in blood is about 3 times greater than that of 1064 nm. This fact can sometimes be put to good use in situations where laser energy must be delivered efficiently through an intervening layer of blood if the Nd:YAG laser being used is capable of operation at 1320 nm.

Melanin

Melanin is the biological pigment that imparts a brown or black color to skin and other soft tissue. It strongly absorbs all visible and ultraviolet wavelengths. Absorption strength increases monotonically (without any significant local maxima or absorption peaks) as wavelength decreases, and increases by more than a factor of 10 between 1000 nm and 400 nm. Melanin absorption is strongest at ultraviolet wavelengths.

Xanthophyll

Xanthophyll is a pigment localized in the neural fiber layer and macular region of the retina (neural retina). It absorbs strongly at a wavelength of 460 nm (a blue-violet color) which is often a consideration in retinal coagulation procedures. In general, laser photocoagulation of the macular region with blue-green wavelengths is avoided for fear of overheating and thermally injuring the neural retina. When macular procedures are performed, a red or yellow wavelength that is not strongly absorbed by xanthophyll is typically used.

Carbon

With the major exception of water, most biological molecules contain carbon atoms. When tissue is heated with a laser in the presence of oxygen (air), some amount of tissue literally burns and leaves a carbon residue, or ash, on the tissue surface. This is commonly referred to by various names, including **carbonization, char, charring, or eschar.**

Tissue charring is usually not desired and measures are taken to minimize its production or remove it once it has been produced. Its presence can induce an objectionable inflammatory response and cause healing complications. Charring may alter the intended laser-tissue interaction in some undesirable way. Carbon particles, which are black, strongly absorb laser energy of any wavelength. When exposed to even low laser power levels, carbon particles heat up rapidly and heat surrounding tissue via heat conduction. This can alter dramatically the intended irradiated volume and alter surgical precision as a result. Tissue heating can be relatively uncontrolled when carbonization is present in significant amounts, especially when using a continuous-wave laser.

Small amounts of surface carbon can produce highly objectionable results when pulsed laser energy is used. Carbon particles can be heated so rapidly that they incandesce and eject explosively from the tissue surface. A transient orange flame and loud acoustic report are typically produced when using pulse energies near 1 Joule and short pulse durations (few hundred microseconds or less). No adverse effects related to such events have been reported that we know about, but such events can be disorienting or objectionable to the surgeon or patient.

Charring can usually be eliminated or greatly reduced by irrigating the tissue surface with water or saline. Irrigation acts to cool the tissue surface and keep oxygen away from heated tissue.

There are situations however where charring is deliberately induced. Visible and near-infrared laser wavelengths are not absorbed strongly enough in some tissue, such as bone or cartilage, for continuous-wave lasers to vaporize and precisely ablate these tissues in a non-contact mode. As a "crutch" of sorts, tissue is heated in the presence of air, and without irrigation, until some amount of char forms on the tissue surface. Once the char is present, it absorbs laser energy very strongly and quickly heats underlying tissue via conduction so tissue can be vaporized with less laser power than would otherwise be required. In this case, charring actually improves the surgical precision of the laser. However, the precision obtained is usually degraded relative to what could be achieved with another laser wavelength absorbed strongly in the uncharred tissue.

Collagen

Collagen fibrils are an important structural component of most types of tissue, including skin, blood vessels, intestine, cartilage, bone, and enamel. Collagen fibrils in soft tissue are thought to unravel in response to mild heating and to play a role in the process by which tissue becomes "sticky" during laser tissue welding procedures. Collagen fibrils expand considerably when heated to the point of denaturation or coagulation. In fact, one of the mechanisms responsible for hemostasis is thought to involve blood vessel constriction caused by expansion of collagen fibrils in the vessel wall. Collagen has strong infrared absorption peaks at 3030 nm, 6060 nm, 6540 nm, and 8060 nm.

Soft tissue

We use the term *soft tissue* in this publication as a catch-all phrase for all fleshy, pliable tissue such as skin, mucosa, and soft internal organs. The term is also intended to imply a significant amount of vascularization and blood perfusion so hemoglobin can be relied upon to heat and vaporize tissue. However, hemoglobin content may vary considerably between different types of soft tissue, resulting in possible differences in absorption and surgical precision when hemoglobin is employed as the main light-absorbing molecule. Soft tissue is also characterized by high contents of water (70% or better) and collagen.

Blood vessels

For our purposes, blood vessels can be thought of as hollow soft tissue. Blood traveling through larger vessels acts to cool the vessel and any nearby soft tissue. This fact can have an important effect on the amount of laser power needed to heat tissue and create a photothermal effect.

The process of cauterization, or **hemostasis**, involves direct heating and coagulation of proteins in blood so that blood becomes viscous or clots and solidifies altogether. Constriction of the vessel wall due to expansion of denatured collagen fibrils may also occur. Laser wavelengths that can penetrate through surrounding soft tissue, and through blood deeply into the vessel lumen itself, seem to be more effective for inducing hemostasis. Thermal heating effects and penetration depth, as controlled by laser wavelength and total laser energy exposure, must typically be adjusted according the size of the tissue's larger blood vessels. In general, higher laser powers and power densities (but below the levels needed to vaporize tissue), longer pulse durations, and/or deeper penetrating laser wavelengths are needed to improve hemostasis.

Temporary constriction of blood vessels in response to heating or other stimulus is called **vasospasm**. Vasospasm is a concern during laser angioplasty procedures to the extent that vasospasm can increase the chances of inadvertent vessel perforation. Some pulsed ultraviolet lasers can induce the opposite effect; namely, temporary relaxation of the vessel wall and lumen expansion.

Cartilage

The main constituents of cartilage are water and collagen. Presumably, cartilage exhibits strong absorption peaks at the same wavelengths as its component materials. About 75% of the total weight of articular cartilage is water, and roughly 70% of its dry weight is contributed by collagen. Most of the remaining weight is that of an organic matrix substance called proteoglycan. Cartilage is generally considered to be avascular and having no significant amount of hemoglobin for mediating laser-tissue interactions.

Bone

Bone is a composite material that consists of an inorganic (mineral) matrix of hydroxyapatite-like calcium phosphate and an organic matrix consisting mostly of collagen. Water is present in both the organic and mineral components. The numbers we have seen suggest that water accounts for 10 to 30% of bone's weight, and hydroxyapatite about 45%. Hydroxyapatite absorbs strongly at 2940 nm and 9260 nm, adding to the strong absorption peaks provided by water and collagen at 1450, 1930, 2940, 3030, 6060, 6540, and 8060 nm. Most of the hydroxyapatite component is probably photodisrupted and removed as fine solid particles when pulsed lasers are used to ablate bone.

Dental tissue

Hard dental tissues such as enamel, dentin, cementum, dental plaque, or caries all contain useful amounts of water, collagen, and hydroxyapatite-like calcium phosphate. The water component is roughly 2 to 3% for enamel and 13% for dentin; the hydroxyapatite content has been estimated at 96% for enamel and 70% for dentin. Presumably, cementum, caries, and dental plaque contain less hydroxyapatite and more water than enamel or dentin, as these materials are typically softer.

Due to strong absorption of 2940 nm laser energy by collagen, hydroxyapatite, and water, erbium lasers (2936 nm or 2790 nm) can precisely ablate all hard dental substances at usefully fast rates, and with relatively low pulse energy and average power. Even healthy tooth enamel, which is highly resistant to visible and other infrared laser wavelengths, can be ablated precisely with erbium laser pulse energy of 100 to 200 millijoules. No topically-applied absorber pigment of any sort is required when using an erbium laser on hard tissue.

Atherosclerotic plaque

Blood vessel blockages, or plaque lesions, consist primarily of fatty deposits (soft plaque), fatty deposits laced with collagen and other fibrils (fibrous plaque), and fatty deposits that contain calcium phosphate (calcified plaque). Plaque is usually white or pale yellow in color and does not absorb visible or near-infrared wavelengths strongly. These wavelengths usually cannot be used to ablate plaque directly. Pulsed mid-infrared and ultraviolet wavelength lasers are already being used to precisely ablate hard and soft plaque. Presumably there is enough water and protein in plaque to allow efficient heating and pulsed ablation at these wavelengths.

Calculus

The terms calculus, or calculi, refer to the stone-like deposits of inorganic salts and organic materials that accumulate in various organs, glands, and ducts. Included are kidney stones, ureter stones, bile duct stones, gallstones, and stones in lacrimal (tear) and salivary glands. Various stone fragmentation, or **lithotripsy**, modalities have evolved that allow stone removal with minimally-invasive techniques. Devices include extra-corporeal shock wave, electrohydraulic, ultrasonic, and laser lithotripters.

Gallstones and bile duct stones tend to be rather soft and crumbly, but are often too large to grasp and crush easily with an endoscopic mechanical instrument. Kidney stones and ureter stones vary from very hard and gravel-like (calcium oxalate monohydrate), to softer stones that are hard to fragment because they lack brittleness (uric acid and cystine), to stones that are soft and crumbly (struvite). Stones come in pigmented and relatively colorless versions, and strongly absorb various laser wavelengths depending on their actual biochemical composition. The 504 nm wavelength used in most pulsed dye laser lithotripters was selected for its strong absorption by most types of calculus, and because it's absorbed weakly by the soft tissue of duct walls.

Medical Laser Delivery Systems

A laser's *delivery system* refers to the physical hardware needed to transfer laser energy from the laser to where it is used to treat tissue. This includes any hardware manipulated by the practitioner to control the application of laser energy at the treatment site. Laser delivery systems also provide some means for enclosing the laser beam as it is guided to the patient so that other personnel working in the room are not exposed to laser energy.

Tissue access and treatment *control* are important issues in the design and use of laser delivery systems. The *distal* or working end of the delivery system must access target tissue in a manner that is safe for the patient and comfortable for the practitioner. Once the distal end of the delivery system is positioned at the treatment site, safety and effectiveness require that laser energy be applied in a controlled fashion to all tissue areas intended for treatment and only to those areas.

Unlike *open* surgical procedures, where a large incision is made to fully expose internal organs so they can be visualized with the naked eye, *minimally-invasive* surgery (MIS) procedures require some special means for visualizing internal tissue and controlling laser treatment inside the body. Laser delivery systems for MIS provide real-time feedback to the surgeon as to what is happening at the surgery site so laser parameters can be adjusted to achieve the desired result.

An optical telescope device called an *endoscope* is frequently used during MIS and is inserted into the patient to relay an image of internal anatomy to an external eyepiece for viewing. Endoscopes are said to provide *direct visualization* of the treatment site. *Indirect* visualization methods provide an "encoded" image of internal anatomy (something other than what one would see with the naked eye or endoscope) and include fluoroscopy (real-time X-ray imaging), ultrasonic imaging, magnetic resonance imaging, and spectroscopic guidance.

Strictly speaking, the term "endoscopic" surgery or therapy implies that an endoscope is used to directly visualize tissue. A broader definition is adopted here to include any procedures performed inside the body using direct or indirect visualization methods. We tend to use the terms "endoscopic" and "minimally-invasive" as synonyms, and refer to endoscopic delivery systems as ones designed to deliver laser energy inside the human body working through a small skin incision or natural body orifice.

To summarize, our definition of a laser delivery system includes 1) apparatus for transferring or guiding laser energy from the laser to the patient (**beam transfer devices**), 2) hardware for controlling the application of laser energy at the treatment site (**beam application devices**), and 3) some direct or indirect method of "seeing" tissue effects created as laser energy is applied (**visualization devices**).

Beam transfer hardware

Devices for guiding laser beams to patients include articulated arms, hollow waveguides, and optical fibers. Optical fibers are often preferred for their relative convenience or economy.

Articulated arms

An **articulated arm** has several "knuckles", or joints, that allow the arm to bend and be configured as necessary to bring the end of the arm to where it is needed. Laser energy travels through and is contained within the arm until energy exits the arm at the distal end. Articulated arms have been used in laser surgery because, until recently, they have been the only practical way to deliver energy from a CO_2 laser to tissue. For reasons that will become apparent below, an articulated arm by itself is not useful for many endoscopic procedures.

Medical Laser Delivery Systems

After exiting the laser enclosure, the laser beam is introduced optically into one end of the first of a series of hollow metal tubes (typically 2, 3, or 4 tubes, each roughly 0.5" to 1" in diameter). The tubes are connected together at their joints in such a way that the angle between adjacent tubes can be adjusted. The laser surgeon adjusts these angles as needed to position the distal end of the last tube where laser energy is needed to treat tissue. The overall reach of the arm is about 4 to 8 feet depending on the arm's design.

A mirror in each knuckle of the arm deflects (reflects) the laser beam and sends it down the center of the next tube to the next mirror in the arm. Unlike the hollow waveguides described below, the laser beam does not strike the inner surfaces of the arm's hollow tubes under normal operating conditions. (The arm's mirrors must be properly *aligned* to prevent this from happening). Once the arm has been aligned at the factory, or upon installation, the mirrors stay aligned as the distal end of the arm is moved around during laser surgery.

Once the laser beam reaches the distal end of the arm, the beam is typically concentrated onto tissue with a ***focusing lens*** attached to the arm in some way. This lens reduces the diameter (and area) of the laser beam on the tissue surface so laser power density is increased to a level that can produce the desired tissue effect. The focusing lens might be incorporated into a handpiece that screws onto the end of the arm, or into a device called a ***micromanipulator*** (see below) used to position the focused laser beam onto tissue more precisely.

Articulated arm delivery systems have a number of problems that limit their utility for MIS procedures. Most arms have large outer diameters (0.5" to 1") over most of their length. Even when arms are terminated with thinner hollow probes (see below) their outer diameters are too big to insert through smaller incisions or natural orifices, especially if other instruments must be inserted simultaneously through the same opening. When articulated arms are used to treat internal tissue, it is usually during an open surgical procedure in which internal organs are exposed with very large incisions for easy access (incisions are big enough for the physician to literally get one or both hands into the body). Nevertheless, articulated arms terminated with hollow probes are used for some endoscopic abdominal

Copyright 2007: JGM Associates, Inc

(laparoscopic) procedures and other MIS procedures where larger and/or multiple incisions are employed (multiple incisions permit the hollow delivery probe to have its own entrance port).

Some endoscopic applications require a delivery system that is thin and flexible at all points along its length. Internal passageways leading to the treatment site can be narrow and tortuous (many bends and angles) such that a delivery system with any large-diameter or rigid sections cannot be used. In some cases, the last inch or so of the delivery system must be bent sharply to properly access target tissue. The tip must be **deflected**, or **angulated**, so the laser beam exiting the fiber can impinge on the target tissue surface at a 90° angle. We do not know of any articulated arm delivery systems capable of distal tip angulation such that the needed angle can be adjusted once the distal end of the probe is positioned within the patient. However, some hollow probe accessories can deflect the beam at the distal tip through some fixed angle to access tissue around corners. A probe with the proper deflection angle must be attached to the arm before insertion into the patient's body.

Other problems with articulated arms apply regardless of whether endoscopic surgery is involved or not. In principle, articulated arms should stay aligned once they have been aligned at the factory. The fact is, however, that normal use misaligns the arm more often than many users would like. When misalignment occurs, the amount of laser energy exiting the arm is much less than what goes in, and the shape of the laser beam coming out is altered. This situation can prevent the laser from being used safely or effectively. Vibrations caused by moving the laser and its articulated arm around between procedures, such as over doorway thresholds and other bumpy surfaces, are often the culprit. A service call is required to realign the arm in many cases.

Another problem related to arm misalignment, but which can have more serious consequences, concerns the use of so-called **aiming beams**. Most medical laser systems employ a low-power aiming laser beam to indicate beforehand where the high-power laser treatment beam will impinge on tissue when the laser's footpedal is pressed. The aiming beam must be aligned properly through the articulated arm so it hits tissue at the same exact spot as the treatment laser beam. Considering the difficulty of maintaining alignment of one beam through the arm, let alone two, it often happens that the aiming beam and treatment beam are not well-overlapped at the tissue surface. Surgical control and patient safety can be compromised if the surgeon is not aware of this situation when it occurs.

Optical fibers

Optical fibers solve the problems associated with articulated arms. The availability of practical optical fibers for high power laser delivery has played a critical role in the growing acceptance of lasers as therapeutic and surgical modalities. The number of new procedures that can be implemented easily, safely, and effectively with a laser will likely increase as fiberoptic laser delivery technology evolves. Optical fiber delivery systems are discussed at considerable length in Section 4. For now, we highlight how fiberoptic systems contrast with articulated arms and provide a general discussion of the role they play in laser surgery.

Optical fibers used in medical laser delivery systems are typically made out of silica or quartz (glass) and have diameters that range anywhere from 0.1 to 1 millimeter (mm). Such thin strands of glass are flexible and are mechanically strong enough to resist breaking. Quartz fibers with diameters of 0.1 to 0.3 mm are very flexible and can be wrapped around a pencil without breaking. Fibers with diameters near 1 mm are much stiffer, but are considerably more flexible and easier to use than an articulated arm. Different fiber diameters are needed depending on the therapeutic procedure being performed and the amount of laser power to be delivered, among other factors.

Optical fibers guide laser energy like a flexible garden hose guides water. Once introduced into one end of the fiber, laser energy is contained within and follows the bends and curves of the fiber until energy reaches the distal end, where it exits the fiber. This ability to guide laser energy, combined with their extreme thinness, flexibility, and mechanical ruggedness, render quartz optical fibers highly useful for endoscopic and other MIS procedures. Thin optical fibers can be inserted easily through millimeter-size skin incisions, through body orifices, or through accessory channels in operating endoscopes. Because they are very flexible, optical fibers can be bent around sharp anatomical corners and with as many bends and curves as necessary to negotiate internal passageways. The distal tip of thinner fibers can usually be deflected or angulated to access target tissue around corners.

Maintenance of optical fiber delivery systems is straightforward since multiple, hard-to-keep-aligned mirrors are eliminated. Although optical misalignment can occur at the fiber input end, the number of optical components is reduced significantly compared to articulated arms and reliability is enhanced accordingly. In fact, misalignment of fiber input optics is rarely a problem no matter how the laser is moved around between procedures. Furthermore, by properly coupling the aiming laser beam into the same fiber used to deliver the treatment beam, it is easy to ensure that the treatment beam and aiming beam are always well-overlapped on tissue. The fiber maintains this spatial overlap "automatically" in some sense. Besides improving convenience during laser surgery, quartz optical fibers have been a boon to endoscopic laser surgery because they are economical to use as **single-use disposables**. Many fiberoptic delivery accessories sell for $100 to $200, allowing them to be used once and then thrown away. Increasingly, in this era of AIDS and lawsuits, it is often preferred that anything that inserts into the patient be thrown away rather than reused.

The "rub" is that not all medically useful lasers can be delivered through quartz optical fibers. In general, laser wavelength dictates whether quartz fibers can be used. Quartz fibers transmit wavelengths between 300 nm and 2400 nm with good efficiency, which means that most of the energy put in one end comes out the other. Transmission efficiencies of 80 to 90% are typical for quartz fibers 2 to 4 meters long. On the other hand, quartz fiber transmission efficiencies for wavelengths longer than 2400 nm, or shorter than 300 nm, are close to zero for long fibers. (These wavelengths are **absorbed** by quartz, which means that laser energy is converted into useless heat by the fiber).

The CO_2 laser, having a wavelength of 10,600 nm, is perhaps the best example of a laser that would be very useful for endoscopic procedures were it not for the fact that quartz absorbs very strongly at this wavelength. Beam energy is completely absorbed after it has traveled only a few millimeters into the fiber. Optical fibers made with an alternative material, called *silver halide*, may eventually prove useful for delivering CO_2 laser energy, but are considered experimental at this time and are not widely used in a clinical setting.

Hollow waveguides

Hollow waveguide delivery systems mimic the performance of optical fibers and can be a practical solution in some situations where optical fibers are desirable but not available. Hollow waveguides are specially-designed hollow metal or plastic tubes that guide laser energy through their internal lumen. Rigid and semi-rigid (or semi-flexible) versions are available. By design, laser energy is reflected off, and guided by, the internal walls of the hollow tube, thereby eliminating the need to maintain alignment of multiple mirrors. Like optical fibers, hollow waveguides maintain good spatial overlap of aiming and laser treatment beams at the tissue surface.

Waveguides are much thinner than articulated arms (diameters of 1.5 to 3 millimeters) and are usually more practical for delivering laser energy inside the body. Articulated arms are frequently terminated with hollow waveguides when performing endoscopic procedures.

The main disadvantages of waveguides (compared to optical fibers) include their relative lack of flexibility and their short overall length. Small-radius and sharp bends must be avoided. A 90-degree bend may cause transmission losses as high as 50%. Although some waveguides are semi-flexible, their use is generally limited to "straight-shot" endoscopic applications in which a skin incision can be made directly over the internal tissue to be treated and intervening organs and tissue can be moved out of the way.

Efficient laser transmission requires that waveguide length be limited to 1.5 meters or less for present hollow waveguide designs. Many waveguide delivery accessories are considerably shorter (12" to 18"). By comparison, optical fibers used in laser surgery are usually 2 to 3 meters long. Such long lengths permit placement of the laser well away from the surgical field, and out of main traffic patterns in the operating room, while still providing usefully long residual lengths for insertion into the patient's body. Waveguides are often attached to an articulated arm to mollify this length problem, thereby incurring some of the problems associated with articulated arms. However, arm misalignment is mitigated by the fact that a shorter arm can be used that has fewer joints and mirrors.

The recent development of "photonic bandgap" hollow waveguides could improve waveguide performance by enabling good transmission efficiency through longer lengths, reducing bending losses, and improving output beam quality. Such waveguides could prove useful at CO_2 laser wavelengths and other wavelengths not compatible with quartz optical fibers.

Beam application devices

Optical fibers and hollow waveguides can be used directly on tissue. However, they are often used along with some other device that may provides added functionality or that controls the application of laser energy more precisely what than can be achieved easily with the unaided human hand.

Fiber handpieces

Since optical fibers are very thin, they can be difficult to manipulate without having them slide through one's fingers, especially when the fiber tip is used in contact with tissue. Accordingly, a handpiece with a diameter approximating that of a pen or pencil is usually attached to the fiber somewhere along its length, usually at a point where the fiber inserts into the patient. The handpiece may also incorporate a rigid, small diameter metal tube (cannula) that actually does insert into the patient, and that has the required combination of bends, straight sections, and angles needed to access target anatomy. The optical fiber inserts into and adopts the shape of the cannula, and the distal fiber tip usually protrudes slightly beyond the end of the cannula. Some handpieces offer a **malleable** sheath or cannula that can be bent to customize it to a particular patient's anatomy.

More sophisticated handpieces provide a mechanism for angulating the distal fiber tip once it has been inserted into the patient. These are referred to as **diverter** handpieces. For example, the last few millimeters of fiber length may be deflected through an angle of 0 to 45 degrees by pulling a "trigger" or turning a knob somewhere on the handpiece. Some method for temporarily locking the handpiece at the desired angle is usually provided. Once laser treatment is completed, the angle is readjusted to 0 degrees so the handpiece can be withdrawn from the patient.

In many laser handpieces, the metal cannula through which the fiber is inserted also provides a channel for suctioning laser-generated smoke or blood away from the surgery site. (The fiber typically does not completely fill the cannula's lumen). Alternatively, the cannula might be used to deliver fluid irrigation to the treatment site. Increasingly, fiber handpieces are being developed that provide suction and irrigation capability along with laser energy delivery, in the same handpiece, so the surgeon can accomplish all of these tasks with one hand. The surgeon's other hand is then free to manipulate an endoscope or other instrument.

Fiber handpieces are offered as disposable items, in which case they are usually permanently attached to the optical fiber, or as reusable items that can be removed from old fibers and attached to new ones. Reusable handpieces are designed to be resterilized by autoclaving or some other conventional method.

In some instances, it may be convenient to use the endoscope itself as the laser handpiece. Most surgical endoscopes have an accessory channel through which an optical fiber can be inserted. Some flexible endoscopes provide the ability to deflect the distal fiber end through some angle.

Hollow waveguides

Hollow waveguides are usually grasped and manipulated directly, and often serve as the handpiece or beam application device for an articulated-arm delivery system. Some of the newer waveguides provide the ability to deflect the laser beam 90 degrees at the waveguide's distal end (a small, fixed mirror is incorporated) or feature a curved-tip design that allows one to treat tissue around corners. Some waveguides have exchangeable disposable tips tailored to specific procedures and anatomical situations. As a result, new hollow waveguide delivery accessories will likely increase the number of situations where carbon dioxide lasers can be used endoscopically.

Although much more useful than articulated arm delivery systems, hollow waveguides are not as convenient to use endoscopically as optical fibers. Waveguides are often considerably larger in diameter at their working tip, and the angle of bent tips cannot be adjusted intra-operatively by the surgeon (angulated) to better access internal tissue. In general, waveguides cannot be used in direct contact with tissue to provide tactile feedback for the surgeon. Waveguides must also be cooled with a flow of gas through the hollow core, which can be inconvenient in some endoscopic situations (bubble formation in fluid media that obscures visualization, risks of complications due to gas insufflation, etc.).

Micromanipulators

A *micromanipulator* is a movable mirror/focusing lens assembly used to control the position of a focused laser beam on tissue with microscopic accuracy and precision. Micromanipulators are typically used with, and attached to, a surgical microscope, so that very small structures can be visualized and selectively treated with very small laser beam spots. Micromanipulators can be used with articulated-arm or fiberoptic beam transfer devices.

The micromanipulator's movable mirror, which controls the position of the focused laser beam on tissue, is controlled with a joystick. The mirror's driving mechanism is "geared down" in such a way as to desensitize, or demagnify, the motion of the laser beam relative to the motion of the surgeon's hand and joystick. For example, the surgeon might move the end of the joystick through a distance of 10 mm resulting in a laser beam movement of only 1 mm on the tissue surface. Precise positional control is afforded that can be much better than that achievable by hand-manipulation of a fiber handpiece or scalpel.

One of the potential benefits of micromanipulator delivery is that surgery can be performed in a completely hands-off manner, and without any hands or handpieces in the way to obscure visualization. This can be a great advantage when treating very small tissue structures during microsurgical procedures. For the most part, use of micromanipulators is limited to open surgical procedures and straight-shot endoscopic applications.

Hollow probes

As we use the term, a **hollow probe** delivery accessory is a short, rigid, tubular extension of an articulated arm. It is usually thinner than the main length of the arm, and is designed to be inserted into the body through an incision or body orifice. Unlike a hollow waveguide, laser energy is not deliberately reflected off the internal surfaces of the hollow probe.

A long-focal-length lens in the "adapter" that connects the hollow probe to the arm is used to focus laser energy onto tissue. The laser beam must be properly aligned through the hollow probe so it does not hit or "clip" probe walls as it proceeds to tissue. Laser power density at tissue, and beam circularity, suffer considerably if clipping does occur, usually to the point where the laser cannot be used safely or effectively. Hollow probe applicators exhibit all of the problems of articulated arms mentioned earlier including difficulty in maintaining alignment between aiming and treatment beams. Hollow probes are used almost exclusively with CO_2 lasers and articulated arms.

Arms vs. waveguides vs. fibers

For laser surgery procedures in which all three delivery system types can be used practically, each offers certain tradeoffs that might be preferred depending on the laser surgeon. For example, articulated arms, hollow waveguides, and optical fibers can all be used with benefit for open or external surgery and for many procedures in the oral cavity. In this subsection, we evaluate each type in terms of tissue access, output beam quality, need for backstops, reusability/disposability (*i.e.*, infection control), and practicality for contact or non-contact use.

Tissue access Fiberoptic accessories are thinner, more flexible, and longer than hollow waveguides, and may therefore offer easier tissue access in some situations. Although all three delivery system types can be equipped with distal tips that treat tissue around corners, only (some) fiberoptic delivery systems give the surgeon the ability to angulate the distal tip after it has been positioned within the patient. Most would agree that fibers provide more convenience compared to hollow waveguides and articulated arms.

However, for some laser practitioners, the advantages of fiberoptics do not always outweigh certain other technical considerations or disadvantages, especially if beam quality is a concern.

Beam quality   Beam quality provided by a laser delivery accessory is another important technical consideration. The term is generally used to refer to the smallest laser spot diameter than can be produced on tissue with the accessory, the roundness (or lack thereof) of the laser spot, and how rapidly the laser spot diameter increases as distance between the end of the delivery accessory and tissue is increased (*i.e.*, how rapidly the beam *diverges* from the end of the delivery accessory).

Articulated-arm delivery systems employ a distal lens to focus laser energy onto tissue. Beam spots as small as 150 microns can be produced when used with CO_2 lasers. Assuming that beam quality of the laser itself is good, lens-focused beams provided by articulated arms are generally very round and have low divergence. When used with CO_2 lasers, laser spot size on tissue is relatively insensitive to arm-to-tissue distance so beam power density on tissue remains fairly constant while treating contoured surfaces (surfaces with peaks and valleys); beam diameter and power density remain constant over a distance of several centimeters in some cases. Low beam divergence reduces the need for the surgeon to readjust arm-to-tissue distance during treatment.

However, the relatively low beam divergence of articulated arm delivery systems frequently necessitates their use with **backstops** to protect normal tissue from inadvertent laser injury. When vaporizing tissue, it is frequently desirable to place a backstop (beam blocking device) between target tissue and normal tissue immediately behind or beneath it, to prevent the beam from affecting normal tissue once overlying target tissue has been removed.

Hollow waveguides typically have a circular cross-section and provide a round beam spot when used in near-contact with tissue. Spot diameters on tissue as small as 150 microns can be produced with some waveguide products. Depending on the quality of the waveguide, and cleanliness of the waveguide tip, beam spots can remain very round as the waveguide tip is moved away from tissue. However, some waveguide products are criticized as having output beams that lose their roundness as the tip is pulled back from the tissue surface, sometimes taking on a jagged, less well-defined shape that can degrade surgical precision and control.

When waveguides are used with CO_2 lasers, waveguide beam divergence is somewhat greater than that obtained with lens-focused articulated arms. This reduces the need for backstops in some situations in that beam power density falls off more rapidly with distance, which can result in less risk of thermal injury to underlying normal tissue. At the same time, higher beam divergence increases the need to readjust tip-to-tissue distance when working on contoured surfaces as needed to maintain constant power density at tissue.

Fiberoptic delivery systems have relatively large output beam divergences, further reducing the need for backstops and increasing the need for adjustment of fiber-to-tissue distance when working on contoured surfaces. Most hot-tip fiber accessories only produce the desired effect when in actual contact with tissue.

Bare (flat-ended) fibers can produce minimum spot diameters roughly as small as the fiber core itself when the fiber tip is used in contact or virtual-contact with tissue (see Section 4). Larger spots are produced as the fiber tip is held farther away from the tissue surface. Typical fiber core sizes range between 200 microns and 600 microns. A very round beam spot on tissue is produced when fiber tips are properly cleaved or polished and maintained free of tissue debris. However, it can happen rather easily that fiber ends lose their polish or cleave during use, and sometimes become chipped, resulting in jagged and less well-defined beam spots.

Infection control Ease and cost of infection control are important considering that the working end of a laser delivery accessory usually comes in contact with patient blood or other body fluids. Some fiberoptic delivery accessories are inexpensive enough (about $100) that the entire fiber can be thrown away after a single use. Some fiberoptic delivery systems have economical fiber tips ($5) that can be used once and thrown away without discarding the entire length of fiber. Other, more expensive fiber accessories can be resterilized and reused. Dental laser fibers can be recleaved and reused many times at a per-use cost of one or a few dollars.

(Many fiber delivery accessories are labeled as single-use items; patient safety can be compromised by improper reuse of fiberoptic accessories).

Some plastic hollow waveguide accessories are inexpensive enough (about $100 to $200) that the entire waveguide can be discarded after a single use. Increasingly however, waveguides are being equipped with economical distal tips that cost $5 to $10 and that are thrown away instead of the entire waveguide. Rigid waveguides are usually designed as reusable devices. Articulated-arm delivery systems typically employ reusable handpieces, but some of the newest arms have economical single-use or limited-reuse handpieces.

Contact vs. non-contact use Contact use of a laser delivery accessory usually minimizes the size of the laser treatment spot on tissue and provides tactile feedback for the surgeon. On the other hand, non-contact use facilitates adjustment of beam spot size on tissue. Large beam spots several millimeters in diameter can be produced easily in a non-contact mode, thereby permitting rapid treatment of larger areas if enough cw or average laser power is available.

Fiberoptic delivery accessories can be used easily and reliably in contact with tissue, or in a non-contact mode. Fiberoptic accessories are highly desirable for many open and endoscopic laser procedures for this reason. Contact and non-contact techniques may be used during the same procedure. The ability of fiber optic accessories to provide the "feel" of a scalpel is a valuable feature for surgeons in many situations.

Hollow waveguides are generally used in a non-contact mode but are sometimes used in contact with tissue. Hollow waveguides employ a flow of CO_2 gas through the central waveguide lumen as a way to keep it cool during use and to keep tissue debris off the inner (optical) surfaces of the waveguide. Contact use interferes with gas flow and may result in waveguide damage. Some users have suggested that hollow waveguides with very small tip diameters, *e.g.*, 350 microns, tend to clog up when used for too long in direct contact with tissue.

Articulated arms are almost never used in direct contact with tissue (at least not in the same sense that fibers are). Presumably, this is because most arm tips have a large tip diameter that would obstruct visualization of the surgery site if used in contact. Most articulated arms employ a flow of CO_2 gas to keep delivery optics clean that would be obstructed if the arm tip were used in contact. Handpieces for articulated arms often incorporate a thin "standoff" that does contact tissue, but doesn't obscure visualization too much. The standoff provides some tactile feedback for the surgeon, but the feel is a poor imitation of that of a scalpel or contact-mode fiber.

Scanning delivery systems

Instead of hand-held laser delivery probes, computerized scanning delivery systems are being used increasingly with CW and quasi-CW lasers for some applications. A computer-controlled mirror located at the end of the beam transfer device (which could be an articulated arm, hollow waveguide, or optical fiber) directs laser energy onto tissue in a highly controlled fashion. Scanners are being used to treat tissue more uniformly, more quickly, or with more precisely defined laser exposure patterns compared to manual methods.

Scanning delivery systems were once used in dermatology for use with some CW lasers to mimic the performance of pulsed dye lasers. Scanned CW lasers seem to be capable of more uniform treatment of vascular skin lesions and benign pigmented lesions than their hand-wielded CW laser counterparts. Clinical results suggest that more uniform treatment with scanners may reduce the chances for hypertrophic scarring when using CW or quasi-CW lasers.

Another area where scanners are having an impact is so-called "char-free" ablation with CO_2 lasers. Tightly focused CO_2 laser beams can vaporize soft tissue with reduced collateral thermal injury, and less charring, compared to use of *conventional* CO_2 lasers with larger beam spots. (Some superpulsed char-free CO_2 lasers can be used without scanners and with large beam spots). However, while tight focusing is acceptable for most cutting applications, vaporization over an area (ablation) is typically done with larger beam spots to increase treatment speed and facilitate uniform ablation. Scanning delivery systems solve this dilemma by enabling rapid and uniform ablation with tightly focused CO_2 laser beams. Automatic beam placement reduces the need for multiple passes over the same area to achieve uniform ablation, further reducing collateral thermal injury and charring. In this way, rapid "char-free" ablation can be achieved with relatively low power CO_2 lasers that aren't superpulsed.

Scanning delivery systems are also used for refractive laser surgery (such as correction of near-sightedness) by permitting rapid and precise ablation of the cornea with focused laser beams in well-defined patterns. The focused beam must be scanned rapidly, and with a very high level of control of beam placement, to produce the desired correction in corneal curvature and refractive power. Some scanners can control beam position with precision on the order of 1 micron.

Visualization devices or modalities

Optical devices for directly visualizing tissue during minimally-invasive surgery include endoscopes and various types of surgical microscopes. Indirect methods include fluoroscopy, magnetic resonance imaging, and ultrasonic imaging. Although present systems do not provide images *per se*, it may be possible to use *spectroscopic* guidance methods during some laser MIS procedures.

Endoscopes

An **endoscope** is a thin, tubular optical device designed to be inserted into the body through a small incision or natural body opening. Its main purpose is to relay images of objects at the inserted end to an external eyepiece for viewing by the physician, with or without magnification.

Virtually all endoscopes use a bundle of very thin optical fibers to transfer light from a lamp source to the distal end of the scope where white light emerges to illuminate internal structures. A **rigid endoscope** uses a series of thick, rod-shaped glass lenses, which cannot be bent or flexed, to relay images back to the eyepiece. Accordingly, all optics are enclosed in a rigid tube that cannot be bent by the physician. Nevertheless, a rigid endoscope can be used to view tissue obliquely; that is, tissue off to one side of the scope rather than directly in front of it. Rigid endoscopes with viewing angles of 0, 30, 70, 90, and 120 degrees are available.

Rather than rod lenses, a *flexible endoscope* employs a flexible fiberoptic imaging bundle to transfer images back to the eyepiece. The entire optical system is encased in a flexible tube that can be bent as needed to negotiate tortuous ducts or vessels. Many flexible endoscopes also provide a steering mechanism that allows the physician to angulate or deflect the scope's distal tip so the viewing angle can be adjusted while the scope is in the patient. Some flexible endoscopes can deflect their tips through more than 90 degrees.

Flexible and rigid endoscopes intended for use during surgery (surgical or operating endoscopes) provide an **accessory channel** for inserting various surgical instruments including laser delivery fibers. The size of the accessory channel depends on the overall diameter of the scope; larger-diameter scopes have larger channels.

In general, rigid endoscopes provide a broader field of view (can view a larger area at any one time), better image quality and visual acuity (can see smaller anatomical details more clearly), and provide brighter images than flexible endoscopes. These performance differences might diminish as flexible endoscope technology evolves.

There are many different types of endoscopes that differ in overall length, diameter, size of accessory channels, and viewing angles depending on the general location of tissue to be visualized. Endoscope outer diameters vary from about 1 millimeter for some ophthalmic endoscopes, to 10 to 12 mm for some **laparoscopes** (endoscopes for intra-abdominal surgery). In any case, such small diameters allow endoscopes to be inserted, and surgery to be performed, through incisions much smaller than those used for open surgical procedures.

Historically, medical endoscopes have been **monocular** devices where only one eye is used for viewing. Depth perception is limited as result. **Stereoscopic** endoscopes are now available that provide much improved depth perception. These are typically used with a 3D display system consisting of a TV monitor viewed through LCD shutter glasses or other TV-based display concept. Some systems use a projection 3D TV display.

Surgical microscopes

A surgical microscope provides a magnified view of tissue so that small structures can be visualized clearly and treated. A large enough working distance is provided (distance between the microscope's objective lens and the tissue surface) to permit unimpeded manipulation of surgical instruments while viewing through the microscope. A typical working distance might be 12 inches. Although they are used mostly for external or open microsurgical procedures, surgical microscopes can also be used to view and operate on internal anatomy to which a direct line of sight can be obtained (*e.g.*, structures in the mouth or throat).

When used in laser procedures, surgical microscopes are often used with micromanipulators (see above) to permit precise positioning of the focused laser beam. Use of a micromanipulator also permits a completely unobstructed view of tissue during laser surgery since laser energy is applied without any hand-held instruments that might get in the way. Specialized surgical microscopes include slit-lamp microscopes and indirect ophthalmoscopes that allow one to visualize intraocular structures looking through the pupil of the eye. Colposcopes are used to visualize tissue in the vagina and cervix.

Other visualization / guidance methods

Other imaging modalities can be used to obtain real-time information regarding laser-induced tissue effects and can be used, in principle, to monitor and control laser treatment procedures. These methods typically provide images that are encoded in some way, and require considerable skill and experience to properly interpret displayed information. Except for fluoroscopy, use of the imaging modalities described below with a laser is experimental or investigational at this time.

Fluoroscopy provides real-time X-ray images of internal anatomy that can be used to position and control laser fiber catheters. Laser angioplasty procedures (surgery inside blood vessels) are typically done under fluoroscopic control because high-image-quality endoscopes small enough to be used inside a blood vessel have not been available (this situation is changing). In general, fluoroscopy is not very useful for visualizing most soft tissue anatomy and exposes the patient to ionizing radiation.

Ultrasonic imaging systems and methods have attracted great interest because they can characterize soft and hard tissue at *or beneath* the tissue surface being visualized. High-frequency sound waves are projected into tissue and reflected off internal structures to create images. Ultrasonic imaging devices can be made small enough to be combined with a laser delivery fiber into a small diameter catheter that can be inserted easily into the body. Such catheters provide the ability to "look" beneath, or on the other side of, tissue surfaces about to be treated with laser energy. It may be possible to monitor and control the depth of laser-induced tissue effects using ultrasonic imaging even in situations where overlying tissue is not vaporized (removed) by laser energy.

Magnetic resonance imaging (MRI) also provides the ability to "see" beneath tissue surfaces and is capable of detecting very subtle changes in soft-tissue anatomy. Deeply penetrating radio waves and magnetic fields are used to construct images. As for fluoroscopic imaging systems, MRI systems require that the patient be positioned beneath a large scanning head of some sort and cannot be incorporated into laser fiber catheters. The use of MRI in conjunction with lasers is experimental at this time, but shows promise for treating unresectable brain tumors and other deep subsurface tissue structures.

Spectroscopic guidance methods use light generated at the tissue surface when irradiated with low power light to control the application of high power laser energy. Laser-induced fluorescence (LIF, see Section 2) signals traveling back through the laser delivery catheter are collected and analyzed by a computer in real time. The information contained in LIF signals can sometimes be used to determine the type of tissue being irradiated, and/or its present physiological state, which in turn can be used to control further application of laser energy. For example, spectroscopic methods have been used to distinguish between different types of atherosclerotic plaque in arteries. One company (KaVo) has developed an erbium laser product that can ablate tartar from root surfaces using spectroscopic guidance to distinguish between tartar and healthy cementum.

Optical Fiber Delivery Systems

This section presents basic optical fiber concepts and defines terminology relating to fiberoptic laser delivery systems.

Bare fiber basics

A laser beam normally travels in a straight line until its direction is changed by some optic such as a mirror. In contrast, optical fibers can make laser light follow curved paths and guide laser energy much like a garden hose guides water to where it is needed. Energy is introduced into the ***proximal*** end of the fiber by reducing laser beam diameter with a lens, or ***focusing*** the beam, so beam diameter matches the small diameter of the optical fiber, so that all or most of the beam energy can enter the fiber. Once coupled into the proximal end, laser energy bounces (reflects) off the fiber's internal optical surfaces and follows the path dictated by the fiber until energy exits the ***distal*** end. The proximal end is attached mechanically as well as optically to the laser. The distal fiber end is positioned near, or in contact with, target tissue intended for laser treatment.

A parameter of key importance for laser power delivery applications is ***fiber transmission***—the ratio of power exiting the distal end of the fiber to that focused into the proximal end. A fiber transmission of 70% to 90% is generally considered adequate for medical laser applications. **Fiber loss mechanisms** that degrade fiber transmission include absorption of laser energy by the fiber material itself, scattering of laser energy out through the sides of the fiber by optical inhomogeneities within the fiber, and bending losses. The material used to make the optical fiber must be chosen to avoid absorption and scattering losses. Fiber manufacturing processes can also affect absorption and scattering within the fiber.

Optical fibers used in laser medicine today are typically made out of silica glass and are referred to as silica or *quartz fibers*. Unless great pains are taken to eliminate water, free water or humidity present during the fiber manufacturing process can result in considerable amounts of OH molecules (ions) in the glass, typically on the order of 1000 ppm (ppm = parts per million). These **high-OH** fibers absorb strongly at many of the same wavelengths as water, thereby prohibiting their use with surgical lasers that emit at wavelengths strongly absorbed by water; e.g., holmium, erbium, and carbon dioxide laser wavelengths. High-OH fibers transmit well at wavelengths between 300 nm and 1150 nm but are usually not practical for use with laser wavelengths outside this range.

So-called **low-OH** quartz fibers are manufactured under reduced water and humidity conditions and typically have OH content in the 1 to 5 ppm range. Low-OH quartz fibers are suitable for the delivery of infrared laser wavelengths as long as 2400 nm. However, even low-OH fibers have too much OH content to efficiently transmit wavelengths longer than 2400 nm through long lengths of fiber. As a result, potentially useful endoscopic lasers such as CO_2 lasers (wavelength = 10,600 nm) and erbium lasers (2790 or 2936 nm) cannot be delivered through commercially available quartz fibers. Low OH fibers tend to be more expensive than their high OH counterparts by roughly 25%.

Bending losses result when a fiber is bent too sharply; that is, when the radius of the bend is too small. Laser energy leaks out through the sides of the fiber at the bend and is not delivered to the distal fiber tip. The bend radius at which large bending losses are incurred depends on the numerical aperture of the fiber (see below) and how light is coupled into the fiber at the proximal end, among other factors.

Bare fibers are used for many laser procedures. The distal end of the fiber is cut and polished, or cleaved with a suitable cleaving instrument, so the fiber endface is flat and perpendicular to the fiber axis. Laser energy emerges from the end of a bare fiber as a solid "cone" of laser energy in that the beam spreads with some cone angle as it travels away from the fiber tip.

Optical Fiber Delivery Systems

In other words, laser energy ***diverges*** from the fiber tip. Using the garden hose analogy, laser energy diverges from a bare optical fiber something like water from a hose equipped with a pistol attachment, with the pistol adjusted to "spray" rather than "stream".

The actual spread or divergence angle of laser energy exiting the fiber depends on the design of the fiber (the fiber's ***numerical aperture,*** or ***NA***), how laser energy is focused into the fiber, fiber length, and degree of fiber bending, among other factors. The numerical aperture of the fiber is determined by the relative difference in refractive index (at the wavelength of light being delivered) between the glass used in the core of the fiber and that of the fiber's cladding material (see below). The actual divergence angle or numerical aperture of light exiting the fiber can be less than the fiber's numerical aperture, but in general is roughly equal to the fiber's design NA.

The diameter of the laser beam emerging from the bare fiber tip increases as the beam travels away from the fiber tip. As an example, the beam emerging from a 0.5-mm-diameter, 0.4 NA fiber (cone angle = 47 degrees) will increase in diameter from roughly 0.5 mm at the fiber tip to a diameter of about 1" at a distance of 1" from the fiber tip. One can easily control beam diameter, area, and power density at the tissue surface by adjusting the distance between the fiber tip and tissue. As described in Section 2, laser power density is often a determining factor for laser-induced tissue effects.

Quartz fibers generally consist of an inner solid core, an outer cladding layer, and a final buffer coating or jacket layer. The ***fiber core*** is the centermost portion of the fiber and is virtually always made out of quartz (rather than plastic) in surgical laser fibers. The fiber core is primarily responsible for carrying and guiding laser energy through the fiber.

The ***fiber cladding*** is a thin layer of material that surrounds the fiber core. The cladding material has a lower "index of refraction" than the fiber core, which allows the process of ***total internal reflection (TIR)*** to occur in the fiber. Total internal reflection at the core/cladding interface causes light to "bounce off the walls" and stay inside the fiber core until it reaches the distal fiber end. The cladding material also serves to protect

Copyright 2007: JGM Associates, Inc

the fiber core from being nicked or scratched, which would dramatically increase chances of fiber breakage when handled. The compositions of the different materials chosen to make the core and cladding determine the fiber numerical aperture or NA.

The cladding material in medical laser fibers can be plastic, in which case fibers are called **plastic-clad silica (PCS)** fibers, or, alternatively, silica glass of slightly different composition than the core material (**glass-clad fibers**). For PCS fibers, the plastic cladding is typically either a soft silicone polymer or a harder fluoropolymer plastic. Fluoropolymer-clad fibers are often referred to as **hard-clad fibers**.

A **buffer coating** is typically deposited over the cladding layer to preserve mechanical strength and improve the durability of the fiber. Buffer coatings may be a polymer material that does not objectionably increase the stiffness of the fiber, such as polyimide resin or silicone, or a very thin coating of metal such as gold or aluminum. A material called Tefzel^{tm} (DuPont), or Nylon, is sometimes used as an outer **jacket** coating to further enhance fiber strength and durability.

Finally, some fiber delivery systems have what is called a **coaxial cooling sheath** as the outermost layer. Unlike the cladding, buffer coating, or jacket layers, the cooling sheath is not bonded directly to inner layers. Instead the sheath is loosely fitted over the jacketed fiber to provide an annular channel through which cooling gas or fluid can be delivered to the fiber tip. Coaxial fiber cooling is sometimes required to prevent buildup of debris on the fiber tip, which can cause overheating and fiber tip damage. Cooling sheaths are usually employed in contact-tipped fiber accessories to guide cooling fluid or gas to the junction of the fiber and contact tip.

Typical core diameters for medical laser fibers range from .050 mm (50 microns) to 1.5 mm (1500 microns). Fibers with core diameters of 600 microns and larger are considerably stiffer than fibers with cores of 400 microns or less. Fibers with 1000-micron cores can tolerate only mild bending without breaking and are useful only for relatively straight-shot endoscopic applications. Quartz fibers with 100 micron cores are about the size of a strand of human hair and are nearly as flexible. A typical glass-clad medical laser fiber might have core-cladding-buffer diameters of 320-350-

500 microns. A coaxial cooling sheath, if present, usually increases the overall diameter to about 1.5 millimeters.

The minimum **mechanical bend radius** of a fiber is a measure of how sharply a fiber can be bent without breaking. It is the radius of the smallest circle the fiber can be bent into without incurring a high probability the fiber will snap. A rule of thumb is that the minimum bend radius for short-term bending is roughly 100 times the diameter of the glass portion of the fiber. The minimum bend radius would be 35 mm in the example cited above (overall glass cladding diameter of 350 microns = 0.35 mm).

Numerical aperture considerations may limit the minimum **optical bend radius** to something larger than the minimum mechanical bend radius. Light leaks out through the sides of the fiber for bend radii smaller than the minimum optical bend radius.

Contact vs. non-contact use of bare fibers

One of the benefits of using a bare-fiber delivery system for laser surgery is that the same fiber can be used easily in a contact or non-contact mode, assuming one is using an appropriate laser wavelength. Switching between tissue effects (vaporization, coagulation, cutting) is accomplished simply by adjusting the distance between the fiber tip and tissue surface and/or by adjusting the laser power setting.

For example, by using higher laser powers with the fiber tip in direct contact with tissue, so that laser spot diameter at tissue is nominally equal to fiber core diameter, one can make fine linear incisions roughly as wide as the fiber itself. This **contact mode** of using fibers is also useful because it provides the surgeon with **tactile feedback** similar to that expected from a scalpel or other mechanical instrument. One vaporizes an area of tissue in a contact mode by moving the fiber in a two-dimensional pattern. One can also coagulate tissue in a contact mode by using a low laser power setting. Some types of lasers cannot be used easily or reliably with a bare fiber in a contact mode.

One switches from a contact to a **near-contact** mode simply by pulling the fiber off the tissue surface by 1 to 2 mm so the irradiated spot diameter is not much larger than the fiber diameter. Because the fiber is away from the surface slightly, the surgeon's view of the treatment site is improved (obstructed less by the fiber tip) which is desirable in some situations. One can cut in a near-contact mode, but with a somewhat wider incision than obtained in direct contact, assuming one is using a strongly absorbed laser wavelength. For some lasers, cutting with a bare fiber in a near-contact mode may result in faster or smoother cutting than when used in contact mode. Precise coagulation may also be accomplished in a near-contact mode with a lower laser power setting.

Non-contact operation of a bare fiber is achieved by holding the fiber tip 3 to 5 mm, or more, away from the tissue surface. One usually intends to treat a large area of tissue quickly when using a fiber in a distant non-contact mode. Since the irradiated spot diameter is quite large (5 to 10 mm, or more) one is afforded only moderate surgical precision and control. Non-contact use of a fiber in a fluid medium requires that the laser wavelength being used not be absorbed strongly by the fluid, since a thick fluid layer intervenes between the fiber tip and the tissue surface. Cutting and vaporization/ablation of soft tissue in a non-contact mode requires a laser wavelength strongly absorbed by tissue, and non-contact cutting/ablation of hard tissue requires strong absorption and high peak laser power.

Direct contact and near-contact fiber methods are typically used when a higher level of surgical precision is required. Assuming that an appropriate (strongly absorbed) laser wavelength is used, smaller irradiated spot diameters at the tissue surface allow one to treat millimeter- or submillimeter-size areas without thermally injuring adjacent tissue. However, this precision is obtained at the expense of treatment speed, so direct contact and near-contact methods are usually practical only for treating smaller areas and relatively small volumes of tissue.

Hot-tip fibers

Lasers with wavelengths that penetrate deeply into soft tissue, including 1064-nm Nd:YAG lasers and 800-nm diode lasers, cannot be used practically to cut or vaporize soft tissue in a non-contact mode. So-called **hot-tip** fiber accessories are almost always used to precisely cut soft tissue with lasers that have deeply penetrating wavelengths. Laser energy is used to heat the tip of the fiber to temperatures well above 100 degrees C so soft tissue can be vaporized or cut by holding the fiber tip in direct contact with tissue. Heat conduction from the fiber tip, rather than direct absorption of laser energy, is the main mechanism by which tissue temperature is raised to the point of vaporization. Hot-tip fibers are necessarily used in a contact mode.

Hot-tip fibers greatly reduce the zone of collateral thermal injury that is produced (greatly improve surgical precision) when using Nd:YAG or surgical diode lasers. Collateral thermal injury zones of 0.5 mm or less are possible in most types of soft tissue.

In laser dentistry, the laser power levels employed when using Nd:YAG or diode lasers are usually in the 3 to 5 watt range. These levels are low enough that bare (flat-ended) fibers can be used safely in a hot-tip contact mode without having to shape the fiber tip in any special way. This fact enables re-use of fiber accessories simply by recleaving the fiber tip. A freshly cleaved bare fiber tip must usually be "broken in" so the tip accumulates a "coating" that allows it to efficiently absorb laser energy and be heated to high temperature.

Cool-tip fibers

Lasers with wavelengths strongly absorbed by tissue can cut and vaporize without the need for hot-tip fiber action. Tissue temperature is raised to high levels by the direct absorption of laser energy. Fiber tips remain relatively cool when using such lasers, and can be used in a contact or non-contact mode with good surgical precision. Pulsed erbium and holmium lasers fall into this category.

Hot-tip vs. cool-tip fibers

The choice between using a hot-tip or a cool-tip laser frequently boils down to a question of surgeon preference or equipment availability. However, without endorsing one or the other, we describe what we understand to be the relative advantages and disadvantages of each. We encourage the interested reader to formulate his/her own opinion by actually using both types of surgical laser modality in an accredited training environment.

Hot-tip and cool-tip lasers can be used reliably in direct contact with soft tissue to precisely cut, vaporize, and/or coagulate soft tissue. Direct contact use of a fiber delivery accessory provides tactile feedback for the surgeon, and allows use of either modality in a wet or flooded surgical field as well as a dry field.

Since hot-tips may remain hot for a second (or so) after the laser has been de-energized, they are accused by some of presenting a safety hazard to normal tissue. Normal tissue can be accidentally touched and thermally injured if the tip is not allowed to cool before moving it. Once the surgeon learns to wait long enough to allow tip cooling, the complaint then becomes, for some users, that waiting is awkward and cumbersome. Hot-tip devices also take some time to heat up which adds further fuel to the "cumbersome" argument. Cool-tip bare fibers do not present such risks or inconveniences.

On the other hand, those who favor hot-tips believe them to be safer than cool-tip devices because most hot-tip fibers produce a thermal effect only when in direct contact with tissue. Cool-tip fibers, also called ***free-beam*** fibers by some, can thermally injure tissue from a distance in a non-contact mode. The point is that normal tissue located behind the targeted structure can be injured inadvertently if laser irradiation isn't halted once target tissue has been removed. So-called ***backstops*** may be needed when using cool-tip lasers but are generally not required when using most hot-tips.

Another issue is consistency of cutting or vaporization from one type of tissue to the next. Hot-tip devices do not rely on direct laser absorption, so cutting and vaporization results are fairly consistent between tissue types. Since cool-tip systems rely on direct absorption of laser energy to heat tissue, variability between tissue types in terms of cutting speed, power requirements, thermal necrosis zones, etc., can occur depending on how strongly tissue absorbs the laser wavelength being used. This is rarely an issue for erbium or holmium lasers since water is present in large amounts in virtually all types of soft tissue. More variability is seen with green-wavelength lasers (frequency-doubled Nd:YAG and argon lasers) since hemoglobin content can vary considerably between different types of soft tissue.

Cool-tip fibers can be used in a contact or a non-contact mode, whereas hot-tip devices must be used in direct contact with tissue. Treatment speed may be reduced for hot-tip fibers in some situations as a result.

Lasers used with hot-tip fibers typically achieve a smooth cutting action on soft tissue similar to that of a scalpel. Lasers used with cool-tip fibers, which may be pulsed erbium or holmium lasers operating at a pulse rate of 30 Hz or less, may cut soft tissue leaving tissue edges that are relatively jagged. Argon lasers (which are cw lasers) may fall somewhere in between in this regard.

Laser beam divergence, focusing, and fiber-coupling

Laser beam **divergence** is the term used to describe how laser beam diameter changes with distance. See Figure 4-1, in which it is assumed the beam is traveling from left to right. For a **diverging** beam, beam diameter increases with distance. The beam divergence value, or **divergence angle**, is the angle subtended by the edges of the laser beam where the beam power drops to near zero. Beam diameter decreases with distance for a **converging**, or **focusing**, beam. A **collimated** beam is one for which beam diameter does not change appreciably with distance. Due to the physical process of light **diffraction**, beams can be collimated only over a finite distance before they transform into a diverging beam.

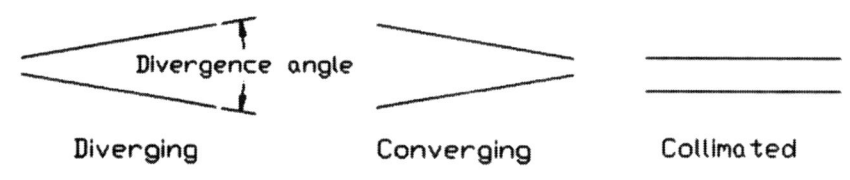

Figure 4-1 *Diverging, converging, and collimated laser beams*

Laser beams are unique in that they can achieve very low divergence angles depending on how the laser is designed and operated. Laser beams that exhibit very good **beam "quality,"** or **TEM00** or **diffraction-limited** beam quality, typically achieve divergence values on the order of 1 milliradian and less. One **milliradian** is one-thousandth of a radian (1 radian = 57.3 degrees). Therefore, a laser beam with 1 milliradian of beam divergence has a divergence angle of .057 degrees. This means that, for every 1 meter of distance the laser beam propagates from the laser, its diameter increases by only 1 mm.

For dental lasers, beam divergence is an important issue when trying to deliver laser beams through an articulated arm or optical fiber delivery system. For articulated arms, beams with adequately low divergence are needed to get the beam through the arm without beam clipping at the arm's internal apertures. When coupling into an optical fiber, low beam divergence is needed to get the laser beam into adequately small-core and small-numerical-aperture fibers.

When a collimated laser beam is focused with a lens, it generally has a converging region just after the lens, a region at the **beam focus** where the beam has a minimum diameter and is collimated over some short distance, and a diverging region located beyond the beam focus. The minimum beam diameter occurs nominally at a distance from the lens equal to one lens **focal length** which is determined by the design of the lens itself. The minimum diameter of the focused spot is determined by the mathematical product of the beam divergence and focal length of the lens being used.

Eqn 4-1 *Focused spot diameter = Beam divergence x Lens focal length*

For the same lens focal length, laser beams with smaller divergence will produce smaller minimum-diameter treatment spots at the beam focus, whereas, for the same beam divergence, shorter focal-length lenses will produce smaller spots at the focus.

When using the focused beam directly on tissue, the diameter of the treatment spot is controlled by controlling the distance between the focusing handpiece (lens) and tissue so tissue is at the focus or somewhere in the converging or diverging region of the beam.

Optical Fiber Delivery Systems

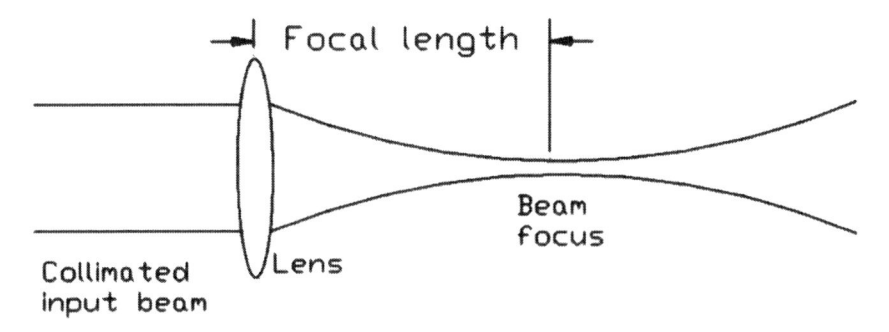

Figure 4-2 *Collimated laser beam focused with a lens*

When coupling the beam into an optical fiber, the proximal end of the fiber is typically placed at the beam focus so that all or most of the beam power can be coupled into the smallest-core fiber possible. In practice, for efficient and reliable fiber coupling, the focused beam diameter is some fraction (*e.g.*, 50%) of the fiber core diameter.

Once the focused beam is coupled into the optical fiber, the beam tries to diverge again as in Figure 4-2. However, total internal reflection (TIR) at the "walls" (core-cladding interface) of the optical fiber contains the diverging light and guides it along a path defined by the fiber.

In addition to a focused beam diameter well-matched to the fiber core diameter, efficient fiber-coupling also requires that the divergence angle of the focused laser beam after the beam focus be equal to or smaller than the fiber's acceptance angle or **numerical aperture** (acceptance NA). The fiber's acceptance NA is equal to the trigonometric sine of the half-angle of the input cone of light, measured in air rather than in the fiber, that can be trapped and guided by the optical fiber. The fiber's acceptance NA is determined by the refractive indices of the materials used to make the fiber's core and cladding structures (see below).

When in air, the divergence angle of the beam after the focus is numerically equal to the convergence angle before the beam focus. (When in the fiber, the beam divergence angle after the focus is reduced by the refractive index value of the glass used to make the fiber core). For most situations of interest, the sine of this convergence angle is nominally equal to the trigonometric tangent, which is equal to one-half the diameter of the

Optical Fiber Delivery Systems

laser beam (at the lens) divided by the lens focal length. The sine of this convergence angle must less than or equal to the fiber NA. In other words, to efficiently couple a laser beam of a given initial diameter into the fiber, the lens focal length can't be too short, otherwise some of the diverging light after the beam focus will leak out through the sides of the fiber.

Accordingly, coupling a laser beam into an optical fiber involves a tradeoff between fiber core diameter and fiber numerical aperture. The beam can be focused to smaller spot and into a smaller-core fiber by using a shorter focal-length lens, but only with the tradeoff of needing a larger fiber numerical aperture to capture all the light. In general, TEM_{00} laser beams are needed for efficient coupling into fibers that have the smallest core diameters and, at the same time, the smallest numerical apertures. For example, the TEM_{00} lasers used in optical communications are coupled into single-mode optical fibers that have core diameters in the 5 to 10 micron range and numerical apertures around 0.1. In contrast, most medical lasers (which, for the most part, are not TEM_{00} lasers) are used with highly multimode optical fibers having core diameters in the 200- to 600-micron range and numerical apertures of 0.2 to 0.4.

Fiber numerical aperture (NA)

Numerical aperture is a quantitative measure of the divergence or cone angle of light that can be efficiently coupled into optical fiber. Numerical Aperture (NA) is defined as the trigonometric sine of the ***maximum*** half-angle of light the fiber can accept or capture, u_{in}, as shown in Figure 4-3.

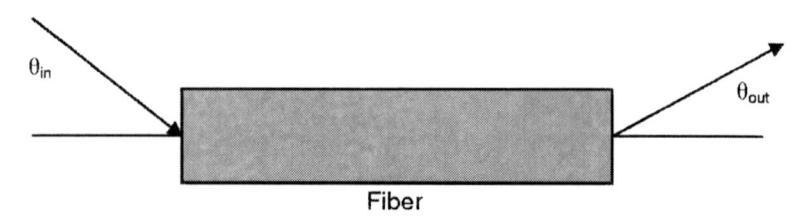

Figure 4-3 *Fiber input and output half-angles*

Optical Fiber Delivery Systems

Fiber NA depends on the design of the fiber itself. For the "step-index" fibers commonly used in laser surgery, the fiber's **acceptance NA** is given by

$$\text{Eqn 4-2} \quad NA = \sin\theta_{in} = \left(n^2(core) - n^2(clad)\right)^{1/2}$$

where u_{in} is the maximum acceptance angle of the fiber, n_{core} is the refractive index of the fiber's core material (at the wavelength of laser light being delivered), and n_{clad} is the refractive index of the fiber's cladding material at the laser wavelength. Refractive indices for the core and cladding materials are in the 1.44 to 1.46 range for quartz fibers, and the difference between the core and cladding refractive indices is usually in the .001 to 0.020 range (core has the higher index). Most surgical or dental laser fibers have an acceptance NA of 0.22 (input half-angle of 12.7°), but some fibers have an NA of 0.39 (22.9° half-angle).

For efficient fiber coupling, the actual input numerical aperture of light injected into the fiber must be less than, or equal to, the fiber acceptance NA. The **input NA** depends on the focal length of the lens used to couple light into the fiber, the diameter of the laser beam at the focusing lens, and, usually to a lesser extent, on how well the laser beam is collimated at the focusing lens. In general, the input NA is designed to be substantially less than the fiber's acceptance NA so there is some margin for alignment error when coupling light into the fiber.

Light goes into the fiber as a solid cone of light with a cone angle equal to two times u_{in}, and emerges from the distal end as a solid cone with cone angle equal to two times u_{out}. However, in some cases in which the input beam is not aligned properly into the fiber, the output light cone may be "hollowed out" so there appears to be a hole in the center of the output beam (usually requires a service call to fix).

If the fiber is short enough, and fiber bending is minimized, the cone angle and numerical aperture of light exiting the fiber at the distal end, or **output NA**, will be equal to the input NA. If the fiber is long enough, or if enough fiber bending is induced, then the observed output NA may be as large as that given by Equation 4-2 even though the input NA is considerably less. Different types of lasers and individual products may exhibit different

Optical Fiber Delivery Systems

output NAs depending not only on the details of the fiber being used, but also on how light is actually coupled into the fiber within the laser head.

The fiber's output NA determines how laser treatment spot diameter changes as distance is varied between the fiber tip and tissue surface. Spot diameter increases more rapidly with fiber-to-tissue distance as output NA increases. Table 4-1 calculates treatment spot diameter versus tip-to-tissue distance for two different fiber output NAs. It shows that, to precisely control treatment spot diameter, fiber-to-tissue distance must be controlled more carefully when using a high-NA fiber.

Table 4-1: Treatment spot diameter versus tip distance and fiber NA

Tip-to-tissue distance (mm)	Treatment spot diameter (mm)	
	NA = 0.22	NA = 0.39
1	.45	.85
2	.90	1.69
3	1.35	2.54
4	1.80	3.38
5	2.25	4.24
10	4.51	8.47

The numerical aperture of the fiber used with a particular laser largely depends on the "beam quality" of the laser. Lasers that exhibit good or high beam quality, including most argon lasers, Nd:YAG lasers, and some erbium lasers, can be used with fiber NAs as small as 0.11 to 0.22. Diode lasers usually have poorer beam quality and may be used with fiber NAs as high as 0.4, depending on output power. However, diode laser technology is improving rapidly such that some newer devices are used with fiber NAs around 0.2.

Some infrared laser fibers, such as sapphire fibers, do not have a cladding and are referred to as **unclad** fibers (even though such fibers are encased in a protective jacket). For these fibers, the "cladding material" is air. Inspection of Equation 1, using an n_{clad} value equal to 1, indicates an NA greater than 1, which is a non-physical result. What this means is that all light coupled into the fiber at one end is guided via total internal reflection within the fiber. However, in practice, light coupled into the fiber at higher input angles experiences much higher loss along the fiber length (due to many more bounces at the fiber wall) so that only light coupled in at lower angles is efficiently delivered through the fiber. The "effective NA" of the fiber refers to that numerical aperture for which light is delivered efficiently, and it depends on the optical quality of the fiber core material, the length of fiber being used, fiber bending, and other factors. Sapphire fibers used in laser dentistry have an effective NA in the 0.2 to 0.3 range.

Infrared (IR) fibers

Quartz optical fibers cannot be used to deliver laser energy at infrared wavelengths longer than about 2400 nm. Carbon dioxide gas lasers (CO_2, wavelength = 10,600 nm) and erbium solid-state lasers (2790 nm or 2940 nm) would be much more useful for endoscopic applications if practical "quartz-like" fiberoptic delivery systems were available. Hollow waveguide accessories are available for erbium and CO_2 lasers that mimic the performance of true optical fibers, and are clinically useful in some endoscopic situations. Hollow waveguides are described further in Section 3.

To summarize their desirable aspects, quartz fibers are a) long (typically 2 to 4 meters, and longer), b) flexible (small bend radii down to 1" and less), c) efficient (80 to 90% transmission through several meters of fiber), d) economical (about $100 end-user purchase price), e) rugged (can withstand substantial mishandling or abuse in a clinical setting without breaking), f) sterilizable without degradation of mechanical or optical performance, and g) have an appropriately long shelf life. One of the most important aspects of quartz fiber performance is that the distal fiber tip can be used in contact with tissue for a sufficiently long period of time, without damaging, and in a wet or dry surgical field. There are no infrared optical fibers that meet all of these requirements at this time. With further R&D, some IR fibers have the potential to meet some or most of them.

CeramOptec (Enfield, CT) has developed a fiberoptic delivery system for CO_2 lasers in conjunction with the General Physics Institute of Moscow. The material used to make the fiber is **polycrystalline silver halide** which transmits laser wavelengths between 4,000 nm and 16,000 nm with reasonably good efficiency (it does not transmit erbium laser wavelengths).

CeramOptec reports that unclad prototype silver halide fibers have delivered 40 watts of CO_2 laser power to tissue with about 70% transmission even with a 1-centimeter-radius bend introduced along the fiber length. Cladded versions can also be made, but typically attenuate about 6 times more than an unclad fiber. Testing results show the fiber to be non-hygroscopic and that it should be possible to use the fiber tip in a wet surgical field. We have not yet seen results of toxicity testing for these fibers.

CeramOptec currently markets unclad silver halide fibers for non-clinical research purposes only. They are available with core diameters of 300, 400, 700, and 1000 microns. Fiber lengths of 1 to 20 meters are available at a price of about $300 to $500 per meter in research quantities (depending on core diameter). According to CeramOptec, there is no reason why silver halide prices can't get down to those of quartz fibers once production volumes reach similar levels. Material components are inexpensive and extrusion processes are straightforward.

Zirconium fluoride fibers have been developed for erbium laser wavelengths. These fibers are available in suitably long lengths, have good flexibility, and transmit efficiently. Core sizes of 100 to 600 microns are available. Current prices for these fibers are about 3 to 5 times that of quartz, but prices are expected to drop as manufacturing volumes increase. Present zirconium fluoride fibers have several drawbacks which may limit their clinical utility: they are not as rugged as quartz fibers (long fiber lengths are prone to breaking in a clinical setting), and the bare fiber tip cannot be used reliably in contact with tissue or in a wet surgical field (the material is slightly hygroscopic and has a fairly low "transition temperature").

Zirconium fluoride fibers could be practical for applications that require only low erbium laser pulse energies and average power levels, and a minimum of handling or bending. Fibers would likely have to be encased in a fairly stiff jacket of some sort. Methods for terminating zirconium fluoride fibers with a contact tip, or a very short length of quartz fiber, are being investigated as ways to permit contact use in a wet surgical field.

Although zirconium fluoride fibers are used clinically, most laser companies considering erbium laser products are looking for alternatives to fluoride fibers. Improved fibers promise to expand the role that Er:YAG and related lasers will play in ophthalmology, dentistry, ENT, and other applications requiring 10W or less average power.

Photran (Amherst, NH) has greatly improved the transmission and throughput capability of its SapphIRe™ *sapphire fibers*, which are made via the company's patented edge-defined film-fed growth (EFG) method. Fiber lengths with good optical quality have been increased substantially. Typical overall transmission is in the 40 to 65% range for 1 to 2 meter lengths of 300-µm-core fiber, including reflection losses of 8% per uncoated endface (total losses of 2 to 4 dB/m). Two-meter-long fibers exhibiting 1 dB/m loss have been demonstrated.

Free-running pulsed Er:YAG input energies of 700 to 800 mJ can be achieved with present sapphire fibers, thereby enabling delivery of as much as 400 mJ to tissue through a 300-µm-core fiber. Photran's testing suggests that delivery of 4 to 5W of average power at 10 Hz is practical. Several dental Er:YAG laser manufacturers now offer sapphire fiber-based delivery accessories for use with their products.

Photran's fibers are grown as unclad single-crystal fibers with an effective numerical aperture of 0.14 to 0.34. A polyester cladding can be applied to the working fiber tip to prevent "evanescent-coupling" and "frustrated-internal-reflection" losses when using the tip in a wet surgical field, although some evanescent absorption occurs in the cladding such that minimizing the cladded length is required. The remaining length can be jacketed with loose-fitting plastic tubing for mechanical protection to prevent scratching and nicking of the fiber. Minimum bend radius is about 6 cm for a 300-µm fiber. Fibers are crystalline and therefore do not cleave in the same fashion as quartz fibers. Polishing with diamond grit is required to achieve a flat and perpendicular fiber endface.

Infrared Fiber Systems (IFS; Silver Springs, MD) now markets its *HP*™ germanium oxide (GeO_2) fiber for high-power erbium laser applications. IFS claims that germanium oxide fibers handle much more power, are mechanically stronger, and are more chemically durable (less hygroscopic) than zirconium fluoride fibers. The fiber material passes agar overlay cytotoxicity and dermal sensitization tests.

IFS's germanium oxide fibers are pulled as clad fibers with a numerical aperture of about 0.2. Bulk fiber losses at 2.94 microns are 0.7 dB/m or less. Fibers cleave with flat and perpendicular endfaces. IFS claims a minimum bend radius of 2.5 cm for a 400-μm-core fiber, and reliable delivery of up to 20W of average power. Fiber core sizes of 100 to 700 microns are available.

Photonic bandgap hollow waveguides

OmniGuide Communications (Cambridge, MA) announced impressive clinical results in which their large-photonic-bandgap hollow waveguide was used to treat recurrent respiratory papillomas. The waveguide was used to deliver about 10 watts of CO_2 laser energy to tissue by inserting the waveguide through the 2-mm-diameter accessory channel of a flexible endoscope.

Photonic bandgap (or photonic crystal) hollow waveguides offer improved performance compared to conventional hollow waveguides, including better transmission through long lengths, reduced bending losses, and better output beam quality from the waveguide. The term "photonic crystal" or "photonic bandgap" refers to the fact that the inner walls of the waveguide are coated with many layers of dielectric material that have precisely controlled composition and thickness. The layers are designed so that light impinging on the waveguide walls with a much broader range of incidence angles can be transmitted through the waveguide with very low loss. OmniGuide claims losses as low as 10% per meter, or 80% transmission through a 2-meter-long waveguide, at CO_2 laser wavelengths.

Even though the core diameter of the photonic-bandgap hollow waveguide is typically much larger than the wavelength of light being delivered, the beam emerging from the waveguide can have low-order-multimode or near-diffraction-limited beam quality (*i.e.*, very low divergence). This is in stark contrast to conventional hollow waveguides and optical fibers used in laser medicine and surgery, which are typically highly multimode optical devices that exhibit much poorer beam quality. This means that smaller treatment spots can be produced on tissue when using longer working distances between the waveguide tip and tissue surface, if so desired, when using the OmniGuide waveguide.

Like conventional hollow waveguides, the OmniGuide device must be cooled with a coaxial flow of nitrogen or argon gas to keep debris from building up at the distal end of the waveguide. The waveguide is terminated with a short metal sleeve that contacts tissue. OmniGuide reports that, in combination with coaxial gas flow, this metal sleeve allows the waveguide tip to be used in direct contact with tissue for extended periods without damage.

OmniGuide expects that their production process will enable very affordable CO_2 laser delivery accessories that may be used as single-use disposable items. Originally designed for making low-loss waveguides for telecom applications, the OmniGuide process can potentially make kilometers of hollow waveguide in a single production run.

Other infrared laser wavelengths can be delivered by changing the thicknesses of the dielectric layers on the waveguide's inner wall.

Argon Lasers

(Sections 5 through 11 were originally written to describe lasers use in dentistry and oral surgery. Many of the same comments made regarding technical laser aspects, and intra-oral use of these lasers, also apply to their use in other areas of the body using endoscopic or minimally-invasive procedures).

Argon lasers are used for gum surgery, curing dental composites during tooth restoration, and tooth whitening. Improved laser tube designs and power-on-demand technology have resulted in longer tube lifetimes and lower operating costs. Argon lasers are being replaced by diode lasers, LEDs, and diode-pumped solid-state lasers.

Technical background—Argon lasers

The active medium in an argon laser is a tube filled with argon gas. The medium is pumped by passing a high-current electrical discharge through the gas (electrical discharge pumping). The discharge ionizes the argon atoms so that electrically charged argon "ions", rather than neutral argon atoms, are responsible for laser emission at blue-green wavelengths. Thus the name "argon ion" laser, which is usually shortened to "argon laser".

The main argon laser wavelengths at which most of the blue-green laser power is provided are 488.0 and 514.5 nm. However, other wavelengths may also be present depending on how the laser is designed. Blue-output lasers intended for dental curing typically generate a mixture of 454.6, 457.9, 465.8, 472.7, 476.5, 488.0, 496.5, and 501.7 nm, without any 514 or 528 nm output. Laser output used for soft-tissue surgery may combine blue wavelengths and the green 514.5 and 528.7 nm wavelengths to maximize output power. Some "green-only" argon lasers generate 514.5 and 528.7 nm, or just 514.5 nm by itself.

Argon lasers are continuous-wave (cw) lasers. Most products can also be operated in a single-pulse or chopped quasi-cw mode. Products that provide a superpulsed emission mode are not available.

Dental argon lasers intended for gum surgery provide a maximum of 2 to 5 watts of blue-green output, and typically require single-phase 208-220 VAC electrical service. These products typically provide a user-selectable "blue-only" output for dental curing. Products intended for dedicated curing applications provide 0.25 to 1W of blue-only output; the higher blue powers are sometimes used for transillumination applications. Curing products that provide a maximum of 0.5 watts of blue-only power operate from a 110 VAC outlet, whereas single-phase 208-220 VAC electrical service is required for curing lasers that provide up to 1 watt of blue for curing and transillumination.

An often cited problem of argon lasers is the tendency of the laser tube to wear out, requiring replacement of the entire tube assembly to repair the laser (rather than just refilling the tube). This can cost as much as $5,000. Depending on the manufacturer, and how heavily the laser is used, tube replacement may be required every 3 to 5 years. Some manufacturers claim substantially longer tube lifetimes.

Although *average* tube lifetimes of 5,000 hours now seem common in the industry, the statistical spread in actual tube lifetimes can be significant. Some tubes may fail in less than 1000 hours, whereas other tubes made by the same manufacturer may have a useful life of 10,000 hours or more. The statistical variation in tube life is at least partly due to manufacturing and quality control measures implemented by the manufacturer; different manufacturers may exhibit substantially different tube lifetime statistics.

Technical background—tissue interactions and effects (Argon lasers)

The argon laser's blue-green wavelengths are absorbed strongly by hemoglobin, which allows this laser to be used via direct absorption of laser energy to cut, vaporize, or superficially coagulate most types of soft tissue. Surgical precision and hemostasis are good in many situations, although precision can be degraded in hemoglobin-poor tissue (more charring and thermal injury to surrounding tissue).

It is often difficult to vaporize hemoglobin-poor tissue in a wet or flooded surgical field, since, in such tissues, blue-green lasers rely on the presence of charring to efficiently heat tissue. (Charring is greatly reduced in a wet surgical field). A hot-tip sculpted quartz fiber may be more useful than a bare cool-tip fiber in such situations, since cutting action is less dependent on tissue pigmentation when using hot-tip fiber devices.

The argon laser's visible wavelengths pass through water and similar clear fluids with little or no attenuation. The laser can therefore be used where a thick intervening layer of saline, saliva, or cerebrospinal fluid (CSF), for example, is present. However, vaporization and cutting of tissue may be prevented in hemoglobin-poor tissue if some amount of charring cannot be produced.

Only argon wavelengths in the 457 to 496 nm range are thought to be photochemically active when curing dental composites and resins. Most of this "blue" output power is provided at 476 and 488 nm. The green (514 nm) wavelength probably serves no useful role in such applications.

Technical background—considerations for intraoral use (Argon lasers)

Argon lasers are used with bare (flat-ended) quartz fibers for contact and non-contact laser procedures on soft tissue. Hot-tip fiber accessories are sometimes used, especially when trying to cut fibrous tissue with low hemoglobin content.

Dental argon lasers are low power devices that provides 5W or less. Fibers must be used in a contact mode or near-contact mode to cut or vaporize soft tissue. This limits the maximum beam spot size that can be achieved at the tissue surface, and, therefore, the utility of these lasers for applications that require rapid vaporization over larger areas.

Historical background (argon lasers)

Low power argon lasers have been used for photocoagulation applications in ophthalmology since the early 1970's. Although higher power products in the 5 to 8W range were introduced in the 1980's for surgical applications, they did not gain widespread acceptance due to lack of adequate power in many instances.

High-power, 15W products were introduced in the latter 1980's, and seem to be more popular for general surgery and related applications. However, these higher power products require tap water cooling and three-phase, 220 VAC electrical service, which severely limits their portability.

Air-cooled argon lasers intended specifically for dentistry and oral surgery were introduced in the early 1990's. Argon lasers for dedicated curing applications appeared in the mid 1990's. Argon lasers intended for tooth whitening appeared commercially in the latter 1990's.

Regulatory status and current clinical uses (Argon lasers)

Dental argon laser products typically have FDA clearances for minor gum surgery, curing of dental composites and resins, and for tooth whitening. Some dental argon laser products also have multi-specialty FDA clearances for ear-nose-and-throat surgery, general surgery, dermatology, and use in other surgical specialties. Not all products have the same FDA clearances.

Recent advances / developments (Argon lasers)

New "power-on-demand" argon laser designs extend tube life by automatically reducing electrical current through the tube when laser output is not being used (when the laser foot pedal is not being pressed). Other innovations that have helped extend tube life include metal-ceramic tube designs, and sealed-mirror designs in which one or both resonator mirrors are sealed directly onto the laser tube.

It would appear that argon lasers have been largely replaced by non-laser light sources (*e.g.*, plasma arc) for curing applications, and by diode lasers and pulsed Nd:YAG lasers for soft-tissue applications. Several companies have discontinued their dental argon laser product lines in recent years.

Editor's note, 2007: While some older products may still be in use clinically, argon laser products are no longer offered commercially by any dental laser manufacturers, as far as we know. Dental argon lasers have been replaced by blue LEDs, diode lasers, and diode-pumped green lasers for curing and tooth whitening applications, and by CO_2, pulsed Nd:YAG, diode, and erbium lasers for applications on soft tissue.

Carbon Dioxide Lasers

Carbon dioxide lasers are used in virtually all areas of laser medicine and surgery, including oral surgery, for precisely cutting or vaporizing soft tissue with hemostasis. Hollow waveguide delivery accessories have improved accessibility of intraoral tissue and overall convenience when using CO_2 lasers. Compact, lower power products intended specifically for use in a dentist's office are available.

Technical background—Carbon dioxide lasers

The active medium in a carbon dioxide laser is a mixture of carbon dioxide, nitrogen, and helium gases. However, only the carbon dioxide molecules participate directly in the stimulated emission process to generate laser energy at an infrared wavelength of 10,600 nm = 10.6 microns (10.6 µm). The active medium is pumped with an electrical discharge, *i.e.*, electrical current is passed through the gas mixture.

At present, all CO_2 laser products intended for dental applications are continuous-wave (cw) lasers. Dental products provide up to 20 watts, but lower power and more economical products are also available. Most products can be operated in a quasi-cw, chopped emission mode for improved surgical control. Some dental products provide a "superpulsed" (higher-peak-power) emission mode to reduce charring when vaporizing soft tissue. Laser emission modes are described in more detail in Section 1.

Carbon dioxide lasers are very efficient compared to most other lasers. Even high power versions that provide up to 100 watts can be operated from a 110 VAC electrical outlet and are air-cooled. The 20 watt lasers typical of dental office products are compact and portable.

Products intended for office use employ a so-called *sealed tube* CO_2 laser design that eliminates the need for gas cylinders and vacuum pumps within the laser. As a result, products are more compact and portable than their flowing-gas counterparts. Carbon dioxide laser products may also be characterized as being *free-space* or *waveguide* lasers, and *DC-excited* or *RF-excited* lasers, as described below.

Sealed tube vs. flowing gas lasers—Carbon dioxide medical laser products are available in flowing gas and sealed-tube versions. In flowing gas lasers, the active gas mixture flows through the electrical discharge region of the laser tube. The gas mixture is constantly being removed from one end of the tube (by a vacuum pump) and simultaneously replenished at the other from an externally- or internally-mounted gas cylinder.

Sealed-tube lasers do not recirculate the gas mixture, do not remove any of the gas mixture, and therefore do not need cylinders for gas replenishment. However, since the gas mixture (laser power) deteriorates with use, the tube must be refilled periodically at the factory with a new charge of laser gas mixture.

Operating costs are comparable, but the numbers we have seen suggest sealed-tube lasers may have considerably lower operating costs compared to heavily utilized flowing gas lasers. We understand that sealed laser tubes must be refurbished roughly every two years at a cost of about $1,500 plus installation. The "operating life" of a gas charge is about 1,000 hours of actual laser use, so that time between refills depends considerably on utilization. A heavily utilized flowing gas laser will go through a tank of gas mixture a week at a per-tank cost of about $100 (roughly $8,000 over two years).

Performance is also comparable. Sealed-tube and flowing gas products provide up to 100 watts of output power. The main performance difference, as we understand the situation, is that flowing gas products typically allow more stable operation at powers less than 5 watts than sealed tube designs, assuming that both types are designed for maximum output of 100 watts. There may also be technical differences in pulse shape when each type of laser is operated in a superpulsed mode (see Section 1) but these differences are rarely considered significant from a user standpoint. Some surgeons like the fact that they can adjust the laser gas mixture in a flowing gas laser, *i.e.*, the relative amounts of helium, carbon dioxide, and nitrogen, in order to maximize pulse energy when operating in superpulsed mode. Others have commented that flowing gas systems can sustain the desired output power level for longer periods of time in superpulsed mode.

The consensus among users seems to be that sealed-tube systems are the future. Sealed-tube and flowing gas lasers are comparable in purchase price, operating costs, and performance. For many users, the deciding factor is convenience. Sealed tube lasers eliminate the hassles associated with ordering and storing gas cylinders, and they are considerably lighter, more compact, and more portable by virtue of elimination of the vacuum pump and gas cylinder. All dental CO_2 laser products described in this section are sealed-tube devices.

Editor's note, 2007 edition: Virtually all office-based medical and dental CO_2 laser products are now sealed-tube devices, and most of these are RF-excited lasers (see below).

Waveguide- vs. free-space-resonator lasers—Standard free-space optical resonators for CO_2 lasers employ a mirror on each end of the gas laser tube (see Section 1). Waveguide resonators employ, as an additional optical element, a short length of hollow waveguide between the mirrors, with the laser gas mixture contained within the hollow waveguide. The "laser mode" within the resonator is defined not only by the laser mirrors but also by the diameter and length of the hollow waveguide. From a user standpoint, waveguide lasers tend to be more compact and electrically efficient for the same laser output power.

Laser output beam quality tends to be somewhat better with a free-space resonator design (*i.e.*, one designed for so-called TEM_{00} operation). This may be an important issue when using an articulated arm delivery system, but is probably not an issue when a hollow waveguide delivery system is used. Newer waveguide lasers have improved beam quality comparable to that of a free-space laser.

RF-excited vs. DC- excited CO_2 lasers—Medical carbon dioxide laser products can also be categorized according to the type of electrical discharge used to excite the laser gas mixture. Although both types operate from a standard 110 VAC outlet, *DC-excited* lasers employ a high-voltage (10,000-20,000 Volts) direct-current power supply (inside the laser enclosure), whereas *RF-excited* lasers employ a low voltage alternating-current internal power supply. In the latter case, discharge current alternates at a very high radio frequency; thus the name "RF".

Internal power supply voltages associated with RF-excited lasers are on the order of 100 volts, which is an obvious safety advantage for technicians who must work inside the laser. RF-excited products are now available that employ free-space resonators or waveguide resonators.

Technical background—tissue interactions and effects (Carbon dioxide lasers)

The CO_2 laser wavelength is absorbed strongly by the water component of tissue. Because water is present in high concentrations, this laser can vaporize or cut all kinds of soft tissue with good surgical precision and reasonably good hemostasis in most situations. Thermal necrosis zones of 100 to 300 microns at cut tissue edges are typical with this laser. CO_2 lasers are particularly useful for vaporizing or cutting dense fibrous (soft) tissue.

The inorganic components of teeth and bone also absorb strongly at the carbon dioxide wavelength. However, because very high temperatures are required to truly vaporize such materials, CW CO_2 lasers cannot ablate or cut calcified tissue without inducing severe charring and other thermal injury to surrounding tissue. Pulsed operation with sufficiently high pulse energy and peak power is required, but such high peak power levels and energies are not available from existing dental CO_2 laser products.

Technical background—considerations for intraoral use (Carbon dioxide lasers)

Carbon dioxide laser energy cannot be delivered through quartz optical fibers. Delivery systems for CO_2 lasers include articulated arms and hollow waveguides (see Section 3). Hollow waveguides and articulated arms intended for intraoral use are now available with disposable straight-tip and contra-angle style handpieces.

Carbon dioxide laser delivery accessories are generally used in a non-contact mode to cut, vaporize, or coagulate soft tissue. Non-contact use is convenient when working on non-fixed movable tissue structures such as buccal and lingual surfaces, and on the floor of the mouth. Non-contact use can also be advantageous when working on ridged or contoured surfaces. In general, contact use of CO_2 laser delivery accessories is not practical.

Historical background (Carbon dioxide lasers)

Invented in the late 1960's, carbon dioxide lasers first appeared as commercial medical products in the early 1970's. They are now used in many areas of laser surgery by virtue of their ability to precisely cut and vaporize soft tissue with hemostasis. Until recently, their applications were limited by articulated-arm delivery systems that were awkward to use at best, and too large in diameter for many potential applications inside the body. High CO_2 laser prices, compared to electrosurgical equipment, probably also contributed to their slow acceptance among surgeons.

New hollow waveguide accessories introduced in the last 20 years or so have enhanced the convenience of using CO_2 lasers in general, and for procedures within the oral cavity in particular. The trend toward endoscopic and other minimally-invasive surgical procedures has created renewed interest in waveguide-delivered CO_2 lasers for some applications. Laser prices have dropped considerably to where one can now purchase a 20-watt CO_2 laser for about $25,000.

Until 1990, most surgical CO_2 laser products were promoted for use in hospitals. A few products were targeted for use in an office setting, *e.g.*, by podiatrists or dermatologists. Products promoted for use by dentists began to appear in the early 1990's once it was demonstrated that a significant dental laser market exists.

Regulatory status and current clinical uses (Carbon dioxide lasers)

Present CO_2 dental laser products have been cleared by FDA only for soft tissue uses in the oral cavity. While hard-tissue applications are being developed, none have gained FDA approval to date. Present soft-tissue applications include gingivectomy, gingivoplasty, frenectomy, and biopsy.

Recent advances (Carbon dioxide lasers)

Rigid and semi-flexible hollow waveguide accessories have enhanced the convenience of using CO_2 lasers and increased the number of possible applications. Waveguide diameters range from about 1.5 to 3.2 mm. Rigid and semi-flexible versions are now available with straight and curved tips. Some waveguide products have interchangeable and disposable tips tailored to specific procedures.

New RF-excited, sealed-tube CO_2 laser technology has made it possible to purchase a 20-watt laser that is very compact and portable. Elimination of vacuum pumps and gas cylinders greatly enhances convenience of use in an office setting.

One company (Lumenis), now offers combination Er:YAG / CO_2 laser products. The Er:YAG laser is intended primarily for hard-tissue applications, whereas the carbon dioxide laser is used primarily for soft-tissue procedures.

Product trends and possible future developments (Carbon dioxide lasers)

Photonic-bandgap hollow waveguides—So-called "photonic-bandgap (PBG)" hollow waveguides have already been introduced commercially for endoscopic CO_2 laser surgical procedures in the trachea and bronchi, and for endonasal procedures (OmniGuide; Cambridge, MA). These high-tech hollow waveguides exhibit reduced bending losses and much better output beam quality compared to conventional CO_2 laser waveguides. It is probably only a matter of time before dental CO_2 lasers are offered with PBG waveguide accessories.

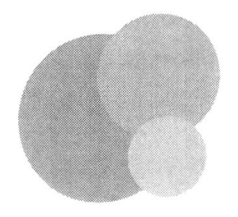

Semiconductor Diode Lasers

High-power 800 nm, 940 nm, and 980 nm diode lasers are now available for minor gum surgery procedures. Diode lasers are very attractive for office-based procedures because they are extremely compact and portable. Diode laser prices should drop over time as technological advances are made.

Technical background—Diode lasers

Like transistors and related electronic components, diode lasers are solid-state devices made out of semiconductor crystals. Diode lasers are similar to the light-emitting diode (LED) display components used in many electronic appliances in that they emit light when electric current passes through them. However, diode lasers emit light that is much more directional, coherent, and monochromatic than LED emission, which allows diode laser light to be focused to small spot sizes as needed to perform laser surgery.

The technical aspects of semiconductor diode lasers are described in Section 1 of this report. Diode lasers used in dental/oral surgery include aluminum gallium arsenide (AlGaAs) lasers that emit at a nominal wavelength of 800 nm, and indium gallium arsenide (InGaAs) devices that emit at a nominal wavelengths of 940 nm and 980 nm. Some dental diode lasers provide up to 20 watts through a 400-micron fiber, whereas others only provide 5 to 10 watts.

All surgical diode laser products now on the market operate from a standard 110 VAC outlet, or are battery-operated. Products are air-cooled, lightweight, and extremely portable. When needed, laser service and repair are usually accomplished by returning the laser to the factory via overnight delivery.

Technical background—tissue interactions and effects (Diode lasers)

Surgical diode lasers are used with hot-tip, contact-mode fiber accessories for precise cutting or vaporization of soft tissue, or with non-contact fibers for deeper coagulation. Clinical data suggest that 800 nm diode lasers provide essentially the same surgical precision and hemostasis as 1064 nm CW Nd:YAG lasers in most situations. Accordingly, surgical diode lasers have found many of the same clinical applications as CW Nd:YAG lasers of comparable output power.

Diode lasers that provide 940 nm or 980 nm wavelength (InGaAs diodes) are relatively new in the field of dentistry. Some manufacturers claim these wavelengths provide better surgical precision in gum tissue than 800 nm, but we have not yet seen any reports of well-controlled scientific studies that support such claims (such studies are reportedly in progress).

Present diode laser products are not capable of the same high peak power levels as pulsed Nd:YAG lasers. As a result, diode lasers are not useful for applications such as first-degree caries removal, for which some pulsed Nd:YAG lasers have gained FDA clearance.

Technical background—considerations for intraoral use (Diode lasers)

Quartz fibers similar to those used with Nd:YAG lasers are used to deliver diode laser energy. Once laser energy is coupled into the fiber, efficient transmission over several meters is possible. However, it is an engineering challenge to efficiently couple diode laser energy into small-core fibers—maximum deliverable power-at-tissue typically decreases as fiber core size decreases. The numerical apertures (NA) of fibers used with surgical diode lasers are usually around 0.4, rather than the 0.2 NA's typical of dental Nd:YAG laser fibers. Advances in diode laser technology will enable increasingly higher power through smaller-core and smaller-NA fibers.

Bare (flat-ended) fibers are used for hot-tip, contact-mode cutting with low power levels of 5 watts and less. Fibers may need to be shaped (sculpted quartz fibers), or terminated with a shaped contact tip (contact-tipped fibers), when using higher power levels in the 10 to 20-watt range, in order to minimize risks of inadvertent thermal injury beyond the intended treatment area.

Historical background and recent advances (Diode lasers)

Advances in high-power diode laser technology throughout the 1980's paved the way for the introduction of the first ophthalmic diode laser products in 1989. These developments included the development of diode lasers that could operate efficiently at room temperature, rather than at cryogenic temperatures, and the development of new diode laser fabrication technologies that allowed more efficient device designs. The result has been a steady upward trend in laser efficiency and output power, and reduction in cost-per-watt of output power. The commercialization of non-medical diode-pumped lasers in the latter 1980's helped increase diode laser production volumes, and reduce prices to where diode lasers could compete with argon lasers in ophthalmic markets.

More recently, surgical diode laser manufacturers have learned how to efficiently couple high power into a single multimode optical fiber as needed for surgical applications. The world's first surgical diode laser, the *Diomed*™ *25* (Diomed; Cambridge, England), was a 25-watt device introduced in 1992. Further improvements in diode laser "array" technology, and fiber-coupling technology, have quickly resulted in higher-power surgical products that can provide 50 to 60 watts through a 600-µm, 0.4 NA fiber. (Some industrial diode laser products can now provide 400W through a 400-micron fiber, or 1000W through a 600- or 800-micron fiber). Surgical diode lasers operating at 940 nm or 980 nm have also been introduced.

The first dental diode laser product, the *Aurora* laser offered by Premier Laser Systems, was introduced in 1995. Since then, numerous other dental diode lasers have been commercialized.

Diode laser prices continue to drop. One can now purchase a 2W diode laser intended primarily for dental hygiene applications for about $9,000, and a 5W laser for gum surgery and tooth whitening procedures for about $12,000.

KaVo recently introduced the first low-power diode laser product (*DiagnoDENT®*) intended for dedicated use as a caries-detection instrument. It employs a red diode laser to induce laser-induced fluorescence (LIF) that identifies the carious lesions.

Sirona recently introduced the first diode laser-based photodynamic disinfection (PDD) system for treatment of periodontal disease. (See Section 2 for a general discussion of photodynamic therapy methods). This device is available in Canada and the EU, but is not yet FDA-approved in the USA as of this writing.

Regulatory status and current clinical uses (Diode lasers)

Most dental diode laser products on the market have FDA clearances for all minor gum surgery procedures in the oral cavity, including subgingival curettage. Some products also have FDA clearance for debridement of soft tissue in root canals that have already been instrumented. However, not all diode laser products have the same FDA clearances. KaVo's *DiagnoDENT®* product is the only one we know of that has explicit FDA clearances for caries detection indications.

Product trends and future developments (Diode lasers)

Higher power through smaller fibers—One can expect diode laser technology evolution to result in more power delivered through smaller-core and smaller-NA fibers, thereby enabling a broader range of surgical applications.

Lower cost-per-watt and diode laser prices—Evolutionary improvements in diode laser component technology will continue to reduce cost-per-watt of diode laser output power. Increasing competition among diode laser component manufacturers will accelerate component price reductions.

FDA approval of Sirona's Periowave—Sirona's *Periowave* diode laser is being used clinically in Canada and the EU for photodynamic treatment of periodontal disease, and FDA approval is pending for use in the USA.

Erbium Lasers

Erbium lasers can precisely cut or ablate all hard dental substances, including healthy enamel, with relatively little pulse energy and average power. Advances in infrared fiber delivery accessories have rendered erbium lasers practical for dental applications.

Technical background—Erbium lasers

Erbium:YAG (Er:YAG) lasers are solid-state devices that use an active medium comprised of a YAG host crystal doped with erbium (Er^{3+}) ions. When in YAG, erbium ions can generate laser emission at a wavelength of 2936 nm. Erbium ions can also be doped into a chromium-sensitized YSGG (yttrium scandium gallium garnet) host crystal, to make a laser medium called erbium, chromium YSGG (Er,Cr:YSGG). When in YSGG, erbium ions can be made to emit at a wavelength of 2780-2790 nm.

Current Er:YAG dental products are flashlamp-pumped lasers that operate in a free-running (FR) pulsed emission mode with pulse durations of 200 to 400 microseconds. Pulse energies of 1J or more can be generated easily, which implies peak laser powers on the order of the 5,000 to 10,000 watts. Average power levels of 20 to 30 watts are possible, but the first dental products to appear are lower power versions with outputs in the 5 to 10 watt range. Erbium lasers are typically air-cooled lasers with an internal water-cooling loop.

For pulse energies of 0.1J to 1J, Er,Cr:YSGG lasers tend to be more efficient than Er:YAG lasers, and, for the same output power, can be somewhat smaller and more compact laser systems as a result. Dental products that employ Er.Cr:YSGG are commercially available.

Unlike Nd:YAG lasers, high power flashlamp-pumped erbium lasers cannot be operated in a continuous-wave (CW) emission mode at room temperature. Diode-pumped CW erbium lasers have been built in the laboratory that operate at room temperature, but only low power devices have been reported.

Erbium lasers can be operated in a Q-switched mode to generate pulses with durations of 100 nanoseconds. Pulse energies of several hundred millijoules are possible. However, such high Q-switched peak power levels and pulse energies cannot be delivered reliably through existing infrared fiber accessories.

Technical background—tissue interactions and effects (Erbium lasers)

Pulsed erbium lasers can cut and ablate tissue with excellent surgical precision (very little collateral thermal injury). The Er:YAG laser wavelength (2936 nm) is absorbed very strongly not only by the water component of tissue, but also by the organic matrix and inorganic hydroxyapatite components that comprise bone, enamel, and other hard biological materials.

The very shallow tissue penetration depths that result from strong absorption, combined with high-peak-power operation, allow erbium lasers to cut and ablate soft tissue with surgical precision comparable to that of cold mechanical instruments. Incisions made in soft tissue with an erbium laser heal almost as quickly as scalpel incisions. Because little collateral thermal injury is produced, relatively little hemostasis is provided in most types of soft tissue when using free-beam erbium laser energy (*i.e.,* when not using a hot-tip, contact-mode fiber accessory).

Shallow penetration depth and high peak power also enable erbium lasers to precisely cut calcified tissue such as tooth enamel, dentin, and bone with virtually no charring. The ablation mechanism for calcified materials is probably a combination of vaporization and photoacoustic processes (spallation) rather than true vaporization. Photoacoustic interactions result in hard tissue being ablated as small microparticles, without having to vaporize the entire hard-tissue mass. As a result, less total energy is required to ablate hard tissue than if photothermal mechanisms were acting alone.

For erbium laser procedures on bone or teeth, the tissue surface is typically irrigated to completely eliminate charring. Even small amounts of carbonization that would normally be of no consequence can result in a loud acoustic report, or "pop", when using high pulse energies. The pop is accompanied by a transient orange flame that emanates from the tissue surface when carbon char is hit by laser pulses. (Carbon particles are rapidly heated to incandescence and eject explosively from the tissue surface). The acoustic report can be disconcerting if the patient is conscious and the bright flash can be disorienting for the dentist. Irrigating the tooth with water prevents charring.

Compared to free-running mode operation, Q-switched operation of the erbium laser improves surgical precision (reduces thermal injury to surrounding tissue). However, the improvement usually does not justify the compromise in laser efficiency incurred by converting to Q-switched operation, or the compromise in fiber reliability that is often involved. If Q-switched products do appear, they will probably be used with articulated-arm delivery accessories.

The 2790 nm wavelength of Er,Cr:YSGG lasers appears to provide surgical capabilities on soft tissue, bone, and teeth that are nominally equivalent to those offered by the 2936 nm Er:YAG wavelength. More research is needed to better define the relative strengths and weaknesses of Er:YAG versus Er,Cr:YSGG lasers for dental procedures.

Technical background—considerations for intraoral use (Erbium lasers)

Er:YAG and Er,Cr:YSGG laser energy cannot be delivered efficiently through quartz optical fibers. So-called "infrared fibers", made of materials other than quartz, are needed that efficiently transmit mid-infrared wavelengths. Such infrared fibers are available commercially for dental applications.

The relatively low pulse rates (1 to 10 Hz) available from some erbium laser products may be considered cumbersome in procedures where dissection of soft tissue is required. Compared to using a continuous-wave laser (equipped with a hot-tip fiber), cutting is often slow or results in jagged, poorly defined tissue edges. This situation might improve as erbium lasers with higher pulse-rate capability become available. Some of the newer dental erbium laser products provide pulse rates to 50 Hz for this reason.

Optical fibers for erbium laser delivery (Erbium lasers)

Infrared fibers for high-energy erbium laser delivery is a technology that continues to evolve. Historically, such infrared fibers have not been as robust, reliable, and cost-effective as quartz fibers used with shorter-wavelength lasers. This situation is changing for the better, but further improvements are probably needed before infrared fibers are considered as reliable and cost-effective as quartz fibers.

Not all dental erbium laser products employ the same kind of infrared fiber material in their fiber delivery accessories. Infrared fiber products from different companies may differ considerably in terms of their reliability.

Historical background (Erbium lasers)

In response to requests from the medical laser research community, Er:YAG laser products were first commercialized in the latter 1980's for scientific research applications. Biomedical researchers such as Dr. Myron Wolbarsht (Duke University, NC) anticipated that Er:YAG lasers might offer unique surgical capabilities based on the fact that water absorbs very strongly at the Er:YAG wavelength. Pre-clinical and clinical research investigations implemented over the last 20 years have laid the foundation for medical and dental erbium laser products now on the market.

Since the introduction by Premier Laser Systems of the first erbium laser product for hard-tissue dental applications in the latter 1990's, other companies such as Biolase, ConBio, and Lumenis have done much to continue improving erbium laser and infrared fiber delivery technology for dental applications.

Recent advances (Erbium lasers)

While the first commercial Er:YAG products were intended for use at pulse rates of 10 Hz or less, some of the newer products are capable of pulse rates to 50 Hz. The development of laser-grade sapphire, zirconium aluminum fluoride, and germanium oxide infrared fibers has stimulated new erbium laser development projects that may result in additional product offerings.

Erbium lasers are now available in "combination" laser systems along with other lasers that provide complementary surgical capabilities. (Combination lasers provide the benefits of owning several different lasers at a lower cost than purchasing separate lasers). Products that combine an Er:YAG laser with an Nd:YAG laser, CO_2 laser, or diode laser are now available commercially.

Regulatory status and current clinical uses (Erbium lasers)

Erbium laser products typically have FDA clearances for soft-tissue and hard-tissue applications in the oral cavity, including caries removal, cavity preparation, and enamel etching. Some erbium laser products now have FDA clearances for indications involving bone. Not all products have the same FDA clearances.

Product trends and future developments (Erbium lasers)

Cost reduction—The benefits of erbium laser dentistry continue to be overshadowed by the relatively high price of such lasers. Manufacturers will continue to strive to reduce the cost of making and selling erbium lasers. Diode-pumped erbium laser technology offers a potential route to lower cost dental erbium lasers.

Fiber delivery improvements—Infrared fiber delivery accessories with improved reliability, and lower cost, will probably be needed for erbium lasers to reach their full potential in dentistry.

New applications—New applications provide a mechanism by which existing products, with existing prices, can achieve deeper penetration into the marketplace. Erbium lasers are being investigated (along with numerous other types of lasers) as a way to modify the surface of dentin and enamel and increase their resistance to caries formation (caries prevention). Caries prevention is a very elegant potential application of lasers in dentistry.

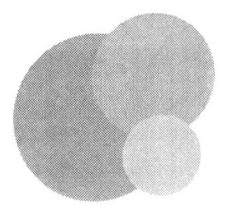

Holmium Lasers

Holmium lasers can cut and vaporize soft tissue with hemostasis, but with the added advantage of laser delivery through very flexible quartz optical fibers. Reasonably good surgical precision and control can be obtained when cutting or vaporizing soft tissue with a bare fiber, in a contact or non-contact mode.

Technical background—Holmium lasers

Holmium laser products are flashlamp-pumped systems. The active laser medium consists of a chromium-sensitized yttrium aluminum garnet host crystal (Cr:YAG) doped with holmium (Ho) and thulium ions (Tm). The chromium ions absorb flashlamp light at blue and green wavelengths and efficiently transfer absorbed energy to holmium ions. The thulium ions act as an intermediary in the energy transfer process (see Section 1 for a discussion of chromium-sensitized laser crystals). The holmium (Ho^{3+}) ions are responsible for generating laser energy at the wavelength of interest, namely 2.1 microns or 2100 nm. This active medium is referred to as Tm,Ho,Cr:YAG, or THC:YAG, and is common to all holmium laser medical products currently on the market.

Products operate in a free-running (FR) pulsed emission mode, typically with pulse durations of 250 to 350 microseconds. Pulse energies up to 5J are used in some (arthroscopic) surgical applications, which implies peak laser powers greater than 10,000 watts. The maximum average power currently available, from lasers used in arthroscopy, is about 80 watts. Dental holmium laser products provide up to 5 watts average power and about 500 mJ of pulse energy. (**Editor's note, 2007:** Dental holmium laser products are not currently offered by any dental laser manufacturers as far as we know).

Holmium lasers are typically air-cooled lasers with a self-contained, internal water-cooling loop. Higher power products have an internal water refrigeration system (chiller), since maximum average power is achieved by maintaining water temperature at or slightly below room temperature. Low-average-power versions, *e.g.*, 1 to 5 watts, do not require an internal chiller.

Holmium lasers can be operated in a Q-switched temporal emission mode to generate pulses with durations of 100 to 200 nanoseconds. However, compared to FR mode operation, output efficiency is typically reduced by roughly a factor of ten. A laser that generates a maximum of 1J in FR mode can generate only about 0.1J in Q-switched mode. Dramatically reduced efficiency, and degraded fiber reliability, renders Q-switched operation of holmium lasers impractical for many surgical applications.

Unlike Nd:YAG lasers, flashlamp-pumped holmium lasers cannot be operated in a continuous-wave (cw) emission mode at room temperature. Holmium lasers have been built in the lab that generate up to 60 watts of cw power, but these devices require cooling of the laser rod to cryogenic temperatures with flowing liquid nitrogen. Such systems are considered impractical for clinical use. Diode-pumped cw holmium lasers have been built in the laboratory that operate at room temperature, and that provide more than 25 watts, but such lasers have not yet been commercialized for any application. More recently, high-power diode-pumped fiber lasers operating at wavelengths in the 2000 nm range, and at room temperature, have been introduced commercially for scientific applications, but none are offered for medical or dental applications as of this writing.

Technical background—tissue interactions and effects (Holmium lasers)

Unlike visible wavelength lasers, and like the CO_2 laser, photothermal interactions with the holmium laser do not rely on hemoglobin or other tissue pigments for efficient heating of tissue. The water component of tissue is responsible for absorbing 2100 nm laser energy and converting it to heat. The 2100 nm absorption depth in water is about 0.3 mm. Since most soft and hard tissues contain significant amounts of water, the penetration depth of 2100 nm is usually quite shallow. When cutting or vaporizing tissue with the holmium laser, actual zones of thermal injury vary from 0.1 to 1 mm depending on exposure parameters and the type of tissue being treated. Good surgical precision and good hemostasis are afforded in many situations. Shallow penetration depths, combined with the high peak power levels characteristic of FR pulsed lasers, allow the holmium laser to ablate hard, calcified tissue. The ablation mechanism for calcified tissue is probably a combination of photothermal and photoacoustic processes (spallation), rather than true vaporization. Calcified tissue is ablated as small solid microparticles, which enables hard-tissue ablation with much less pulse energy than if total vaporization were required. (Holmium lasers are not FDA-approved for any dental hard-tissue applications).

For holmium procedures on bone or hard dental tissue, the tissue surface is typically irrigated to minimize charring. Small amounts of carbonization that would normally be of little concern can result in a loud acoustic report when using pulse energies of 1 joule or more. This acoustic pop is sometimes accompanied by a transient orange flame that emanates from the tissue surface (carbon particles are probably heated rapidly to incandescence, and eject explosively from the tissue surface). To date, this situation has not resulted in any patient or surgeon injury that we know of, but it is generally avoided by irrigating the tissue surface during or in between laser exposures. The acoustic report can be disconcerting to the patient if the patient is conscious, and the bright flash can be disorienting for the surgeon. Low pulse energies near 0.1 J seem to alleviate this problem, but may not be enough energy to quickly perform some procedures.

Hemostasis when cutting or vaporizing with the holmium laser is acceptable to good in most types of soft tissue. Hemostasis appears to be better than that of a CO_2 laser, but not as good as a free-beam CW Nd:YAG laser (1064 nm).

Compared to free-running mode operation, Q-switched operation of holmium lasers improves surgical precision (reduces thermal injury to surrounding tissue). However, the improvement usually does not justify the compromise in laser efficiency incurred by converting to Q-switched operation, or the associated compromise in fiber reliability.

Technical background—considerations for intraoral use (Holmium lasers)

Efficient delivery of holmium laser energy through several-meter lengths of optical fiber requires use of "low-OH" quartz fibers. These are quartz fibers for which special measures have been taken to eliminate water and humidity during the fiber manufacturing process (see Section 4).

The pulse durations characteristic of free-running holmium lasers are long enough for delivery of large pulse energies through small-core fibers. Pulse energies of several joules can be delivered reliably through fibers with core diameters as small as 100 microns. Considerably more energy can be delivered through larger core fibers. In contrast, one can reliably deliver only about 0.05 to 0.1 J (50 to 100 mJ) through a 300-micron-core fiber when in Q-switched mode.

Holmium lasers can be used reliably with bare fibers in either a contact or a non-contact mode. Their high peak power provides a "self-cleaning action" that prevents fiber damage (due to overheating) that can result from debris build-up on the fiber tip. A portion of the energy in each laser pulse literally blows debris off the distal fiber endface, so debris build-up is prevented. As a result, bare (flat-ended) fiber delivery accessories can be used reliably for contact cutting or vaporization, or for non-contact vaporization or coagulation of soft tissue. No coaxial fiber cooling is required.

Because holmium laser fibers can be used in direct contact with tissue, they can be used in a wet or completely submerged surgical field in spite of the fact that the 2100 nm wavelength is strongly absorbed by water. The intervening water layer that might be present is usually too thin to significantly attenuate pulse energy before it reaches the tissue surface, or it is vaporized out of the way by the first part of the laser pulse.

The relatively low pulse rates (10 Hz) available from dental holmium laser products may be considered cumbersome in procedures where more extensive tissue dissection is required. Compared to using a continuous-wave laser, or a 100-Hz pulsed laser equipped with a hot-tip fiber, non-contact cutting with the holmium laser often results in jagged-edge incisions and may be considered too slow for some applications. Dental holmium lasers are probably more useful for procedures which require relatively little cutting and dissection and which involve non-contact vaporization of soft tissue over an area. (Larger spot diameters can be used for non-contact vaporization, enabling reasonably fast treatment even at 10 to 20 Hz pulse rates).

The acoustic impact of joule-level pulsed laser energy on tissue can eject blood, mucus, or saliva on the tissue surface into the air. Protective eyewear and face masks may be needed to reduce biohazards for the practitioner and room personnel.

Historical background and recent advances (Holmium lasers)

Until 1987 or so, holmium lasers were liquid-nitrogen-cooled devices that were confined mostly to research laboratories. Although some surgical laser R&D was done at holmium wavelengths, the inconvenience and complexity of flowing liquid-nitrogen systems discouraged laser companies from offering a commercial medical product. Around 1985, research conducted primarily in the Soviet Union resulted in chromium-sensitized host materials that paved the way for the development of efficient room-temperature holmium lasers. Schwartz Electro-Optics, Inc. (SEO; Orlando, FL) was the first to adopt this new technology in the USA and introduce a commercial scientific laser product. SEO's *Laser 1-2-3* was introduced in 1987 and is intended for surgical laser research at 2100 nm and other infrared wavelengths. This product has done much to stimulate R&D regarding the use of holmium and other new solid-state infrared lasers for surgical applications.

Recent advances in holmium laser technology include improved laser output efficiency and increased average power and pulse rate capability.

Products intended for arthroscopic surgery can provide 80 to 100 watts at pulse rates of 30 to 50 Hz.

The first (and only) dental holmium laser product was introduced by Excel-Quantronix in the early 1990's. This product had limited commercial success and was discontinued in the latter 1990's. (The holmium laser apparently suffered from the fact that it was not as useful on soft-tissue as a hot-tip Nd:YAG or diode laser, and was not as good as Er:YAG on hard tissue). At present, no holmium laser dental products are commercially available as far as we know.

Light-Emitting Diodes (LEDs)

Many high-power blue LED sources are now commercially available for dental curing applications. Like diode lasers, LEDs are very attractive for office-based procedures because they are extremely compact and portable. LED prices should drop significantly over time as technological advances are made.

Technical background—LEDs

Like transistors and related electronic components, light-emitting diodes (LEDs) used in curing light sources are solid-state devices made out of semiconductor crystals. They are similar to the LED display components used in many consumer electronic appliances in that they emit light when electric current passes through them. However, curing LEDs typically provide much higher power levels (*e.g.*, 5W), and specifically at blue wavelengths compatible with camphorquinone (CQ) and 1-phenyl-1,2-propanedion (PPD) photoinitiators.

LEDs are similar to diode lasers except that LED semiconductor chips are fabricated without the resonator micro-mirrors that diode laser chips have. This reduces cost compared to diode lasers, but LED beam quality is typically much worse than that of a diode laser. LED beams can't be focused to the same small spot diameters as diode laser beams.

Unlike lamp-based curing lights (halogen and plasma arc), LEDs are typically not replaceable by the user. However, LEDs have very long lifetimes such that frequent replacement is not necessary.

Higher-power blue LEDs often require a fan in the handpiece if the light source is to be used for extended operation. (Devices that don't have a fan usually have an automatic shutdown feature if the LED gets too hot due to extensive repeated use). Fans usually result in a more noisy handpiece, but some designs claim to have "whisper-quiet" fans to prevent overheating.

All curing LEDs now on the market operate from a standard 110 VAC outlet (or 220 VAC) or are battery-operated. Products are air-cooled, lightweight, and extremely portable. When needed, laser service and repair are usually accomplished by returning the laser to the factory via overnight delivery.

Technical background—material interactions and effects (LEDs)

Blue LEDs are typically made of gallium nitride (GaN) semiconductor material and provide wavelengths in the 400 to 490 nm range. Shorter wavelengths in the 400 to 420 nm range are thought best for PPD-based materials, whereas 440 to 490 nm wavelengths are typically needed for CQ-based materials.

The latest blue LED sources provide intensities in the 1000 to 2000 mW/cm^2 range, which is high enough to cure most resins and composites fast enough to be competitive with halogen light sources. Products offer a selection of tips which provide different treatment spot diameters and, therefore, different maximum curing intensities. Tips are available for interproximal curing as well as curing spots as large as 10 mm.

Some LED curing sources provide multi-function capability including large-spot curing, interproximal curing, tacking, and transillumination (looking for cracks, caries, etc.) capabilities.

LED-based light sources are now being used in diagnostic applications such as detection of oral cancer (LED Dental; *VELscope*®). These applications use the LED to stimulate laser-induced fluorescence (LIF; see Section 2) in tissue, which is analyzed to assess whether tissue is normal or not.

Light-Emitting Diodes (LEDs)

Technical background—considerations for intraoral use (LEDs)

LED beam quality is good enough for efficient coupling of light into multi-fiber bundles with diameters in the 10-mm range. When smaller curing spot diameters are needed, the multi-fiber bundle typically tapers from an initial diameter of 10 or 11 mm down to the desired treatment spot diameter.

Many of the newer curing LED products have a slim handpiece design (pen-style or wand-style, as opposed to gun-style) and a 360° rotating angulated tip, thereby greatly improving access to all areas of the mouth.

Historical background and recent advances (LEDs)

High-power blue LEDs were first introduced for curing applications by Premier Laser Systems (Irvine, CA; no longer in business) in the latter 1990's or early 2000's. However, blue LEDs with enough power to compete with other curing light sources have become available only in the last three or four years.

LED Dental (Vancouver, BC) received FDA and Health Canada approvals for its *VELscope* product in 2006. This device uses a narrowband blue LED to excite natural tissue fluorescence which is analyzed using optical filters to distinguish between healthy and abnormal tissue.

Regulatory status and current clinical uses (LEDs)

Most blue LED products have FDA clearances for curing of dental resins and composites. Some products also have clearances for tooth whitening, but, in general, curing light sources are not designed for the continuous or longer-term exposures used in tooth whitening procedures.

Product trends and future developments (LEDs)

Higher power, more wavelengths, lower cost—One can expect high-power LED technology to evolve quickly. Higher power levels will be provided by lower-cost LED components. Curing devices that provide near-UV and blue wavelengths in the 350 to 400 nm range, and perhaps in a user-selectable fashion, will become more common.

Other applications—Oral tissue is relatively easy to access and amenable to a wide range of possible light-based diagnostic and therapeutic procedures in which an LED light source is feasible to use.

Free-Running, Pulsed Nd:YAG Lasers (1064 nm)

Pulsed free-running (FR) Nd:YAG lasers are used primarily for soft-tissue oral surgery, but the relatively high peak powers provided by these lasers enable some applications on teeth. Whereas the first products introduced in the early 1990's provided a maximum of 3 watts at tissue, present pulsed Nd:YAG dental lasers typically deliver 5 to 6 watts for faster treatments. Fiberoptic delivery of laser energy enables numerous soft-tissue management procedures.

Technical background—Pulsed Nd:YAG lasers

Like other Nd:YAG lasers, the active medium in free-running (FR) pulsed Nd:YAG lasers consists of an yttrium aluminum garnet (YAG) host crystal doped with neodymium (Nd^{3+}) ions. The Nd ions are responsible for generating laser emission at any one of three medically interesting wavelengths: 1064 nm, 1320/1340 nm, or 1444 nm. Current FR Nd:YAG dental products available in the USA provide only the 1064 nm wavelength.

Free-running Nd:YAG lasers differ from CW Nd:YAG lasers in terms of peak power and pulse durations generated. When CW Nd:YAG lasers are operated in a true CW mode, average power and peak laser power are the same and are limited to maximum values of about 100 watts (as medical products go). Some CW Nd:YAG products can be operated in a quasi-cw "superpulsed" mode, but even then peak powers are less than 300 to 400 watts, and pulse durations are usually much longer than 1 millisecond.

In contrast, FR pulsed Nd:YAG lasers typically emit with pulse durations in the 0.1 to 1 millisecond range (100 to 1000 microseconds) and can easily achieve peak power levels of 1,000 to 10,000 watts. Nd:YAG products that operate in free-running pulsed mode (usually) cannot also be operated in a cw emission mode. Commercial dental products provide maximum pulse energies of 200 to 800 mJ depending on the product.

Present generation FR Nd:YAG medical products are flashlamp-pumped lasers. Diode-pumped FR Nd:YAG lasers have appeared as research products, but are probably still too expensive to compete effectively in the dental laser marketplace.

FR Nd:YAG dental lasers are typically air-cooled (no external tap water required) but have an internal water-cooling loop. Present dental products are limited to 25 watts or less; most dental products provide less than 10 watts. Dental products usually operate from a 110 VAC electrical outlet.

Technical background—tissue interactions and effects (Pulsed Nd:YAG lasers)

FR Nd:YAG lasers are used mostly for photothermal laser-tissue interactions, *i.e.*, to heat tissue. When used to cut or vaporize soft oral tissue, FR Nd:YAG lasers are used with hot-tip, contact-mode fiber accessories to achieve good surgical precision. Low-power versions (5 to 6 watts or less) are used with bare (flat-ended) hot-tip fibers for contact cutting and vaporization of soft tissue. Fibers are used in a non-contact mode only to coagulate soft tissue at lower power levels, or to provide hemostasis over an area. When using higher average powers (10 to 20 watts), shaped fibers or contact tips, rather than flat-ended fibers, may be needed to limit thermal necrosis zones to acceptable levels.

Compared to 1064 nm, the 1320 nm and 1444 nm Nd:YAG wavelengths are absorbed much more strongly by the water component of tissue. Penetration depths in soft tissue are reduced accordingly, which in turn provides enhanced surgical precision when using fibers in a non-contact mode. So far, these alternative Nd:YAG laser wavelengths have not gained widespread popularity among surgical laser practitioners. (A 1340-nm dental laser product was once available in Europe, but is no longer available as far as we know).

Technical background—considerations for intraoral use (Pulsed Nd:YAG lasers)

The 1064 nm wavelength penetrates deeply (several millimeters) into most types of soft and hard tissue. This situation presents potential risks of thermal injury to pulp, periodontal ligament, or bone when working on or near teeth. However, actual risks depend critically on average laser power, pulse energy, degree of focusing/defocusing at the fiber tip (*e.g.*, whether or not a shaped fiber tip is employed), and whether or not active tissue cooling is used, among other exposure parameters.

Most of the clinical dental experience gathered to date has been with low power products, *i.e.*, 3 to 6 watts and less, used with bare (flat-ended) fibers. Higher-power FR Nd:YAG products provide faster cutting and vaporization, but may require sculpted fibers or contact tips to maintain adequate surgical precision during gum surgery procedures.

According to dentists who use them, low-power FR Nd:YAG lasers are sometimes used to provide laser-induced analgesia of teeth without injections of local anesthetic. The mechanism behind laser-induced analgesia is not well understood. Such applications have not been approved by FDA.

High-peak-power operation allows FR Nd:YAG lasers to ablate small amounts of hard dental tissue such as carious enamel or dentin. Relatively low absorption of the 1064 nm wavelength makes it difficult to ablate healthy enamel with pulse energies of 500 mJ or less. Enamel etching typically requires application of a topical absorber to the enamel surface. Some pulsed Nd:YAG products now have FDA clearance for removal of first-degree caries, but no other hard-tissue clearances have been granted by FDA.

The relatively long pulse durations of FR Nd:YAG lasers enable reliable delivery of laser energy through small-core optical fibers. Efficient delivery of 1320 nm or 1444 nm Nd:YAG wavelengths requires low-OH quartz fibers, whereas 1064 nm laser energy can be delivered through low-OH or high-OH quartz fibers.

Historical background and recent advances (Pulsed Nd:YAG lasers)

Free-running pulsed Nd:YAG lasers had been used for industrial applications, such as metal welding, for more than a decade before they were first used medically. The advent of CW Nd:YAG lasers used with hot-tip fiber accessories, in the mid-1980's, laid the foundation for use of FR Nd:YAG dental lasers with hot-tip fiber accessories on soft-tissue.

Original research by TD Myers and WD Myers in the mid-1980's showed that FR Nd:YAG lasers could be used with convenient fiberoptic delivery accessories not only for soft tissue surgery, but also for some applications on teeth. Subsequent clinical research in Europe and Canada, in the late 1980's, set the stage for FDA-clearance of the first FR Nd:YAG dental laser products in the early 1990's, but only for soft-tissue applications.

Early FR Nd:YAG dental products provided only 3 watts of average power and were limited to pulse rates of 30 Hz or less. Present-generation products provide 5 to 10 watts and maximum pulse rates of 100 to 200 Hz. As a result, these newer lasers enable considerably faster contact cutting and vaporization when using hot-tip fiber accessories. Pulse rates of 100 Hz or greater seem to be advantageous for achieving good hot-tip action (less time for the fiber tip to cool between pulses).

Regulatory status and current clinical uses (Pulsed Nd:YAG lasers)

Pulsed Nd:YAG laser dental products typically have FDA clearance for hemostasis, incision, excision, and vaporization of soft tissue in the oral cavity, and sulcular debridement / laser curettage. Most pulsed Nd:YAG products now also have FDA clearances for "selective" removal of first-degree enamel caries, in patients of all ages. (Caries must be confined to enamel). One company, Millennium Dental Technologies, recently gained the first explicit FDA clearance for their Laser Assisted New Attachment Procedure (*Laser-ANAP®*) which has been shown to result in new attachment of periodontal ligament to root surfaces. Not all products have the same FDA clearances.

Product trends and future developments (Pulsed Nd:YAG lasers)

Pulsed Nd:YAG vs. diode lasers for soft-tissue surgery—Now that surgical diode lasers have been introduced in dentistry, such lasers may eventually replace pulsed Nd:YAG lasers for many soft-tissue oral surgery procedures. As of this writing, pulsed Nd:YAG lasers can deliver more average power through 200- and 300-µm-core fibers than diode lasers, which may be an important advantage for some soft-tissue applications. Another reason pulsed Nd:YAG lasers may continue to be used for soft-tissue oral surgery is that they can be combined easily (and affordably) with other pulsed lasers, such as an Er:YAG laser, in the same enclosure.

Pulsed Nd:YAG lasers may be preferred to cw lasers for some soft-tissue applications, to the extent that high-peak-power emission results in reduced collateral thermal injury, even when using hot-tip fibers (as is claimed by some manufacturers). However, it remains to be documented, and become widely accepted, that pulsed Nd:YAG lasers offer significant advantages compared to CW Nd:YAG or diode lasers used with hot-tip fibers, when treating soft tissue.

Smaller pulsed Nd:YAG lasers—The emergence of ultra-compact diode lasers has prodded manufacturers of pulsed Nd:YAG lasers to reduce size and weight, and to a lesser extent, the price of their pulsed Nd:YAG products. Further reductions in Nd:YAG laser size and weight will be enabled by advances in the relevant laser component technology. Diode-pumped pulsed Nd:YAG dental lasers are probably still several years away, but will eventually become feasible from a cost standpoint (they are already technically feasible).

Lasers vs. Electrosurgical Instruments

This appendix is intended for information and educational purposes only. We do not intend to endorse or recommend the use of lasers over electrosurgical modalities for any procedure, or vice-versa.

Electrosurgical (ES) cutting and coagulation instruments can often be used instead of a laser, and usually at considerably reduced cost. Continuing technological improvements in electrosurgical instruments are making these devices more useful for endoscopic applications. In this section, we compare lasers and electrosurgical modalities along a number of dimensions that might be considered important.

Laser modalities

Laser modalities for conventional soft-tissue surgery (cutting, vaporization, coagulation) include CW Nd:YAG lasers, diode lasers, frequency-doubled Nd:YAG lasers, erbium lasers, holmium lasers, carbon dioxide lasers, and argon lasers. All of these can be used with flexible optical fiber delivery systems, except carbon dioxide (CO_2) lasers, which are used with articulated arms and hollow waveguide delivery systems. All of these lasers can be used to precisely coagulate tissue, or to precisely vaporize or cut soft tissue with hemostasis.

Electrosurgical modalities

Monopolar ***active electrodes*** include loops, scissors, knives, and snares, and can be used to cut tissue with hemostasis or to coagulate tissue. Monopolar active electrodes are used in conjunction with a broad-area ***dispersive***

(ground) electrode through which electrical current is returned to the electrosurgical generator. The dispersive electrode is usually placed on the patient well away from the site where surgery is performed (with the active electrode), and must be in intimate electrical contact with the patient if dispersive electrode burns are to be avoided.

Bipolar devices incorporate two closely spaced electrodes on the same handpiece, eliminating the need for a separate dispersive electrode. Bipolar ES devices are used primarily for coagulation tasks that require a higher level of surgical precision (see below) than is normally provided by a monopolar instrument. Bipolar probes for cutting tissue are available, but we understand these are prone to "clogging up" between the electrodes with charred tissue, and are often considered not practical for many cutting applications.

Performance issues

Performance issues include surgical precision and control, tactile feedback, ability to perform multiple surgical tasks without having to exchange probes, compatibility with physiologic irrigation or fluid distension media, neuromuscular stimulation, and compatibility with endoscopic instrumentation and techniques.

Surgical precision

A key issue in most surgical applications is **surgical precision**. In essence, the term refers to the thickness of the zone of thermal necrosis at cut tissue edges when cutting or ablating (removing tissue over an area) with a surgical instrument. Hand-operated (as opposed to motorized) mechanical instruments induce no thermal necrosis, and are the gold standard for surgical precision. One necessarily trades surgical precision for hemostasis capability when selecting among surgical modalities, since hemostasis depends on the zone of thermal necrosis at separated tissue edges.

The actual zone of coagulated tissue obtained when using a specific thermal modality varies according to operating parameters, technique, type of tissue, etc. Nevertheless, it is possible to characterize various modalities in terms of the surgical precision they afford, on average, considering how each is typically used. The reader should note that both laser and ES devices continue to evolve technologically, so that today's observations may have to be modified in the future.

In general, lasers are considered more precise surgical instruments than ES devices. Thermal necrosis of several millimeters or more is typical when using monopolar ES instrumentation. Bipolar devices appear able to achieve thermal zones of about 1 mm, and perhaps less in some types of tissue. Laser thermal zones typically fall in the 0.05 to 1 mm range and can approach zero for some types of lasers (excimer lasers and erbium lasers).

A related issue is **surgical control**, which is a term we use to refer to a modality's ability to deliberately remove layer by thin layer of tissue. It is closely related to surgical precision in that the thinnest layer that can be removed, or that one would want to remove, is usually proportional to the minimum thermal necrosis zone that can be achieved. As an example of the surgical control that can be achieved, lasers used for correcting refractive disorders can remove corneal tissue in layers that are less than 1 micron thick.

Cutting, coagulation, and vaporization

Surgical tasks performed with lasers and ES devices include vaporization (ablation), cutting (incision, excision, or dissection), spot coagulation, coaptive coagulation, and broad-area coagulation. Lasers and ES devices are evaluated in terms of their ability to perform each of these tasks with a minimum amount of switching between different hand-held probes during a procedure. In the ideal case, one would like to use a single instrument to perform any of these surgical tasks, which would reduce operating time and instrumentation costs.

Vaporization or ablation refers to tissue removal over a two-dimensional area. Lasers are considered useful for precisely vaporizing tissue, whereas ES devices are considered much less useful. In most vaporization applications, the thermal necrosis zones obtained with a monopolar ES device are too large, or the vaporization process is too slow. Electrosurgical devices vaporize tissue by holding the small-area point of the instrument in direct or virtual contact with tissue, so that only very small areas can be vaporized simultaneously. Tissue tends to stick to the ES probe, further reducing vaporization speed. In contrast, some lasers can quickly vaporize tissue in a non-contact mode, simultaneously vaporizing millimeter- or centimeter-size areas. Tissue sticking is not a problem when laser fibers are used in a non-contact mode, but can be a problem for some types of fiber delivery systems used in a contact mode.

Cutting is a special case of vaporization in which a narrow linear area of tissue is vaporized to separate tissue edges. **Incision** usually refers to a single long or short cut. **Excision** implies several incisions as needed to remove a volume of tissue *en bloc,* rather than trying to vaporize the entire tissue mass. **Dissection** implies repeated incisions to separate tissue, usually to separate an organ or other large mass from its "bed" in the body.

Although monopolar ES devices, and to a lesser extent bipolar devices, are useful for cutting, it is our impression they are most useful in situations where relatively little cutting or dissection is required, such as cutting of vessels, ducts, or bands of tissue. Accumulated thermal necrosis can be extensive when multiple repeated incisions are made in the same general area. Tissue tends to stick to the ES device, resulting in slow cutting speeds and increased operating times where extensive dissection is required. Charring of tissue is usually significant, which can obscure tissue cutting planes. Although lasers leave something to be desired in terms of speed when extensive cutting is required (compared to mechanical instruments), they often cut with less charring, with reduced thermal necrosis at cut edges, and with less sticking of tissue to the cutting probe compared to ES instruments.

As far as we can determine, **spot coagulation** is the forte of ES modalities. Monopolar and bipolar devices can be used to coagulate large blood vessels and stop bleeding in situations where most lasers are useless. For the most part, these are situations where the bleeding vessel is large and easily identifiable. This large-vessel hemostasis capability of ES devices goes hand-in-hand with the relatively poor surgical precision they afford compared to lasers. Bipolar devices are useful for more precise spot coagulation and hemostasis of smaller blood vessels. Lasers provide even more precise spot coagulation capability.

Coagulation of tissue over a substantial area of tissue (**broad-area coagulation**) is useful in situations where bleeding is diffuse (does not originate from one or a few easily identified vessels). In many cases, bleeding occurs from a bed of very many capillary vessels (oozing) or simultaneously from numerous relatively small feeder vessels. Lasers used in a non-contact mode, and with larger spot sizes on tissue, tend to be more useful than ES methods in such situations. The laser beam can be used easily and quickly to "paint" the bleeding area to effect hemostasis. One can use some monopolar ES devices in a non-contact mode to simultaneously coagulate an area of tissue (or **fulgurate**) but precision and control are usually so poor so that fulguration is of limited utility. Furthermore, it is frequently recommended that ES devices *always* be used in direct or virtual contact with tissue in the interest of patient safety.

In general, the ability of any hemostasis (coagulation) modality to quickly stop bleeding from large vessels is enhanced if the hemostatic probe can be used to compress tissue while it is being heated (**coaptive coagulation**). Mechanical compression acts to coapt the sides of the vessel so it can be sealed off with less energy input to tissue, and in less time. Coaptation frequently increases the diameter of the largest blood vessel that can be stopped with a given modality. Bipolar ES probes (and heater probes) are particularly useful for coaptive coagulation. Laser fibers are often considerably less useful, either because they are too thin at their tip or because one cannot reliably exert enough mechanical force to compress vessels without breaking the tip. However, some sapphire-tipped fiber delivery systems have been developed (laser bipolar dissector) that may prove useful for coaptive coagulation of large vessels and ducts using lasers.

Electrosurgical methods can involve considerable switching between probe types when changing the desired surgical task during a procedure. Monopolar knives, loops, forceps, etc. must be interchanged depending on the desired surgical objective. One may also have to switch between monopolar and bipolar devices depending on the surgical precision required. In general, the fact that ES devices are best used in direct contact with tissue increases the need to switch instruments.

Because most of the lasers mentioned earlier can be used in a contact or a non-contact fiber mode, the need to change instruments during a procedure may be reduced considerably. Lasers used with bare (flat-ended) fiber delivery systems, and that have a strongly absorbed laser wavelength in tissue, seem to be particularly advantageous in this regard.

At least to some extent, the ability to tailor surgical treatment to specific anatomical situations is compromised by the requirement that a single hand-held probe be used. It is not likely that a single hand-held probe will ever be suitable for all surgical tasks and anatomy. Thus, the ability to easily switch between probes during a procedure can be a decided advantage in some situations. Our point is that once a probe has been chosen for a particular anatomical situation, one would like to use that same probe for as many surgical tasks as possible during the intended treatment procedure.

Compatibility with irrigation methods

Irrigation of tissue with water or saline is useful in many surgical situations. Wetting the tissue surface tends to cool the surface and keep oxygen away from the surface. As a result, irrigation frequently reduces thermal necrosis, charring, and smoke production when using thermal modalities like lasers and ES devices. Irrigation is also useful for clearing blood from tissue surfaces so that bleeding vessels can be located and cauterized. The use of some devices for automatically cleaning endoscope tips (without removing the scope from the patient) may result in inadvertent irrigation of tissue. Finally, fluid distention of organs and joints, as employed to facilitate some endoscopic procedures, requires surgery in the presence of water, saline, or other water-based fluid media.

The use of electrosurgical devices is not compatible with physiologic, electrolytic fluid media (fluids that contain dissolved ions to preserve osmotic balance in tissue, and therefore conduct electricity). Electrolytic irrigation media can increase the risks of alternate site ES burns (see below) due to uncontrolled current return pathways. As a result, when ES devices are used with irrigation, the irrigation medium is usually pure, low conductivity water. However, pure water is not considered a physiologic medium in many situations and it can upset the normal osmotic balance in tissue. Patient morbidity secondary to abnormal water absorption by tissue may be increased. This issue is particularly important for arthroscopic applications, where Ringer's lactate solution is usually the preferred fluid distension medium.

Since electricity is not involved in the tissue interaction process, laser surgery methods are inherently compatible with physiologic irrigation and fluid distension techniques. Lasers that can be used with quartz optical fiber delivery systems are particularly advantageous in this regard.

Neuromuscular stimulation

The reliance of ES devices on electricity for induced tissue effects is also an issue as far as unwanted stimulation of muscles and nerves (***neuromuscular stimulation***). Muscles can jerk violently upon application of an ES device. This can make it difficult to maintain contact with tissue and is another reason why ES devices are often more useful for spot-type applications than they are for procedures that require repeated or prolonged contact with tissue. In the worst case, neuromuscular stimulation can result in inadvertent thermal injury to normal tissue resulting from accidental touching of tissue with the ES instrument during violent movement of target tissue in unexpected ways. In general, laser methods usually do not stimulate involuntary muscle contraction.

Tactile feedback

Surgeons often rely heavily on tactile feedback from their instruments during surgery. Surgeons also like to use the same probe to "palpate" tissue (touch tissue without treating or injuring it) without having to switch to a separate palpation probe. Electrosurgical probes are used in direct contact with tissue and thus also provide tactile feedback during surgery. Some mechanically blunt ES probes can be used to palpate tissue if one is careful not to activate the footswitch during palpation.

Lack of tactile feedback was a common argument against using lasers in the earlier days of laser surgery (up until the mid-1980's). Today, fiberoptic laser delivery systems can be used with the fiber tip in direct contact with tissue, thereby providing tactile feedback, or in a non-contact mode that does not. In many instances, the same fiber tip can be used to palpate tissue without mechanical or thermal injury to tissue assuming the laser is not accidentally activated during palpation. So-called "hot-tip" fiber devices must be allowed to cool after laser use before trying to palpate tissue (see Section 4).

Compatibility with endoscopic instrumentation and techniques

Important issues regarding endoscopic compatibility are physical size, compatibility with irrigation and aspiration methods, and ability to deflect the working tip under endoscopic control to properly access tissue. Compatibility with irrigation and aspiration methods was addressed earlier. Fiberoptic laser instruments with deflectable tips are currently more common and available than ES devices with deflectable tips, as far as we know.

Perhaps the biggest issue is physical size. In general, one likes thinner instrumentation for endoscopic procedures so that smaller skin incisions can be used. Diameters of existing fiberoptic laser probe accessories are in the 0.5 to 2 mm range. Standard monopolar and bipolar ES devices are typically millimeters (*e.g.*, 2 to 4 mm) in diameter at their working tip, although small diameter bipolar ES instruments are available for microsurgical procedures.

An issue that is closely related to physical size has to do with the size of the area that can be treated simultaneously, *i.e.*, without having to move or reposition the probe tip. ES modalities suffer from a common requirement that they be used in direct contact with tissue. This requirement limits the area that can be treated simultaneously with a thin device, and therefore treatment speed (operating time is increased). In contrast, lasers can be used with thin fiberoptic probes that allow easy access to internal tissue through small incisions, and at the same time enable treatment in a non-contact mode with treatment spot diameters considerably larger than the fiber diameter. This allows simultaneous treatment of several square millimeters or centimeters, if so desired, and if enough laser power is available. Operating time can be reduced significantly as a result. Fiberoptic laser delivery systems that must be used in contact with tissue at all times also suffer the same compromise as ES devices (access through smaller incisions *versus* treatment speed).

Safety issues

Lasers and ES devices are thermal modalities capable of igniting combustible materials, especially in an oxygen-rich insufflation environment. Appropriate caution must be exercised at all times. In general, use of an oxygen-rich or other flame promoting atmosphere is to be avoided.

Lasers present more of a potential ocular hazard to the surgeon, patient, and OR personnel than ES devices. Protective eyewear is required for all operating room personnel during laser-based surgical procedures, which is not the case when ES devices are being used. Interestingly, surgeons have sometimes received burns around the eye as a result of arcing from ES instrumentation to a surgical endoscope.

Lasers used to vaporize tissue with long-focal-length lens delivery systems frequently require the use of "backstops" to reduce risks of injury to tissue located behind the targeted tissue structure. Normal tissue is likely to be vaporized, or otherwise thermally injured, if the laser is not deactivated as soon as target tissue in front of the normal tissue is removed. Lasers used with bare (flat-ended) fiberoptic delivery systems involve similar, but lessened, risks (because laser energy diverges more rapidly beyond the target structure) so that backstops are required less often. Most types of hot-tip fiber delivery systems vaporize only when in direct contact with tissue and present even less risk to back-lying normal tissue. Conventional monopolar ES devices can, in some situations, "arc" through millimeters of air to thermally injure tissue not directly contacted with the instrument.

Some laser delivery systems require use of a coaxial flow of gas for cooling purposes. Use of such delivery systems for endoscopic procedures (and some open procedures) can increase the risks of air-gas embolism, pneumothorax, subcutaneous emphysema, or gas distension pain for the patient. The need for coaxial gas cooling can usually be avoided by choice of a proper laser and fiberoptic delivery system. Similar considerations apply to some uses of ES modalities, where gas flushing of the probe or tissue may be required.

Electrosurgical devices are often accused of electrically interfering with patient monitors during surgery. The dispersive electrode of the ES unit must be located well away from the electrodes of any patient-connected monitoring device in order to minimize electrical interference. Use of ES modalities on patients with pacemakers is frequently avoided for this reason. A well-designed laser does not present electrical interference problems.

A big potential safety issue regarding ES devices has to do with electrically induced burns at patient sites other than the intended site. These have been categorized as dispersive electrode burns, active electrode burns, and alternate site burns (see the ECRI reference at the end of this appendix).

Dispersive electrode burns result when the dispersive electrode is not in intimate electrical contact with the patient at all times during ES use, or when a dispersive electrode is used that is too small in area for the electrical power levels being used. "Dispersive electrode contact monitors" are available that are designed to reduce the risks of dispersive electrode burns. Their use is advocated by hospital consulting firms like ECRI (Emergency Care Research Institute; Plymouth Meeting, PA). There are no risks associated with the use of well-designed lasers, that we know of, that might be considered analogous to dispersive electrode burns.

Active electrode burns result from over-treatment (excessive use of power) and activation of the ES unit when the ES probe is inadvertently touching tissue not intended for treatment. Similar risks apply to laser usage.

Alternate-site burns have occurred where electrical insulation on cabling, the handpiece, or the handpiece shaft has either come off or was not in place originally. For some types of monopolar probes, insulation is not generally applied all the way to the working tip of the device. This has been a problem primarily for endoscopic procedures where inadvertent tissue contact near, but not at, the working tip is more likely. Newer ES probes intended specifically for endoscopic use are more thoroughly insulated along the shaft length.

Alternate-site burns can also occur via "leakage currents" that are coupled through perfectly good insulation to non-insulated metallic devices. The burns occur where the non-insulated device contacts tissue. Leakage currents are a consideration primarily when the ES device is used in close proximity to the non-insulated device, as can occur during endoscopic procedures (leakage current can couple into an endoscope with a non-insulated metallic eyepiece and shaft). Alternate site burns can also result from electrical current returning to ground via pathways other than through the dispersive electrode, such as through temperature probes. Burns result from the fact that the patient contact area of the unintended ground electrode is much smaller than that of the intended dispersive electrode.

The reader should note that, when considering endoscopic or laparoscopic procedures, one should distinguish between conventional ES units and ES units specifically designed for laparoscopic use (see references 4 and 5). The use of conventional ES equipment has been recognized as potentially hazardous for laparoscopic use by the American Association of Laparoscopic Gynecologists (Ref. 4). Conventional ES refers to unipolar equipment with voltages of 1200 Volts or more and bipolar devices with peak-to-peak voltages of 600 Volts or more.

Ergonomic issues

Although there are many ergonomic aspects that could be analyzed, we concentrate on portability and start-up time as these often have an important impact on device utilization. Equipment that can be moved easily between rooms and buildings, and used in various operating suites, is more likely to be utilized by multiple specialties. Equipment that can made be ready to use quickly is more likely to be used in emergency situations and other instances where usage is not scheduled beforehand.

As we understand them, most ES units are small and compact, and plug into standard 120 VAC outlets. A dedicated outlet is not required in most instances. Units are typically air-cooled, obviating the need for tap-water cooling services in the room. Our impression is that most of the setup time involved relates to making sure the site for the dispersive electrode is prepared properly, and that the electrode is properly contacted to the patient.

Most (non-diode-laser) surgical lasers operate from a single-phase 208 VAC outlet. Non-diode surgical lasers that do operate from a 120 VAC outlet often require a dedicated outlet. Most of the newer products are air-cooled rather than water-cooled. Some surgical laser products have a warm-up time of 5 minutes or so, although many products have virtually no warm-up time. For lasers that don't have a warm-up time, most of the setup time involved relates to connecting the optical fiber delivery system to the laser, calibrating it, and draping it into the surgical field.

Diode-based lasers, including direct-diode lasers and diode-pumped solid-state lasers, are dramatically changing the ergonomics and economics of using lasers. Diode-based laser products are typically very small and compact, extremely portable, and much more reliable than non-diode-based lasers. They can usually operate from any standard 110 VAC outlet, and some can operate from a battery. In the future, diode-based lasers may be able to compete with ES devices not only in terms of performance, but also in terms of price and per-use cost.

Cost issues

Lasers are usually much more expensive to purchase and amortize (per procedure) than ES equipment. We understand that a typical purchase price for an ES unit is $3,000 to $5,000. High-end systems may cost as much as $20,000.

Surgical laser prices vary from $10,000 at the low end, for a low-power surgical diode laser, to as much as $150,000 for a frequency-doubled Nd:YAG laser or high-power holmium laser. Prices for diode-based laser products are expected to drop significantly over time as prices for high-power diode components come down.

Reusable probes for ES units cost about $300 to $700 and can be reused many times. Per-use cost is rarely an issue as we understand the situation. The purchase prices we have seen for single-use, disposable ES probes are in the $30 to $50 range. In contrast, the per-use costs of most fiberoptic delivery system accessories are in the $100 to $600 range, if used only once. However, sapphire-tipped contact delivery systems can be reused as many as 5 to 10 times, bringing per-use costs down to about $60 to $70 in the best case ($150 on average). Some bare-fiber delivery systems can achieve per-use costs in the $5 to $10 range, if users are willing to "recleave" and reuse fibers. Institutional policy may prohibit reuse of fiber accessories, as most are labeled by the manufacturer as "single-use only" items.

Much more so than ES devices, some laser products may have limited FDA clearance or approval regarding the variety and types of surgical procedures in which they can be used. This can affect device utilization. The FDA does not regulate how and when surgeons can use lasers, but use of a laser for procedures not sanctioned by FDA can have an impact on liability litigation. Most, but not all, CW Nd:YAG laser and surgical diode laser products have broad FDA clearances that span all or most surgical specialties, for cutting, vaporization, or coagulation of soft tissue. The same can now be said for surgical (15W or higher) frequency-doubled Nd:YAG, argon, erbium, and holmium laser products. Note that FDA clearances and approvals are granted on a product-by-product basis; manufacturers should be contacted directly regarding FDA clearances for their specific products.

Reimbursement issues

The issue of third-party reimbursement for laser procedures is in a state of flux. There are relatively few CPT and ICD-9 codes specifically for laser-based procedures, so the process of getting reimbursed for the full cost of laser procedures is something of an art. Getting fully reimbursed is frequently an uncertain proposition. On the other hand, as we understand current practices, getting reimbursed for costs of using ES devices is rarely an issue.

Laparoscopic cholecystectomy (endoscopic gallbladder removal) is a good example of what can happen. In some states, insurance carriers have decided they will definitely not reimburse for the added costs associated with using a laser for the procedure, instead of an ES modality.

References

1. See "Electrosurgical Unit" in *Encyclopedia of Medical Devices and Instrumentation*, Vol. 2, pp. 1180-1203, Webster JG, ed., John Wiley & Sons (New York, 1988).
2. See "Update: Controlling the risks of electrosurgery", *Health Devices* Vol. 18, No.12, pp. 430-432, December 1989. Published by ECRI (Plymouth Meeting, PA).
3. Hauser K, "Laser vs. electrosurgery: advantages and disadvantages", *Medical Electronics*, February 1990, p. 167.
4. Hauser K, "Laser vs. electrosurgery in laparoscopic cholecystectomy", *Medical Electronics*, December 1990, p. 90.
5. Hauser K, "Electrosurgery: macro vs. micro", *Medical Electronics*, June 1985.
6. See "Are general surgeons ignoring lessons of gynecology?", *Clinical Laser Monthly*, January 1991, p. 1.

CPSIA information can be obtained at www.ICGtesting.com
Printed in the USA
BVOW08s2243260214

346144BV00001B/89/P